A Text Book of

TRAFFIC ENGINEERING

Semester – V

FOR

THIRD YEAR DIPLOMA COURSE IN CIVIL ENGINEERING GROUP

As Per MSBTE's 'I' Scheme Syllabus

VAIBHAO K. SONARKAR

B.E. (Civil), M.E. (Env. Engg.)
V.B.V. Polytechnic
Vasai Road (W),
Dist. : Palghar

Dr. DINESH KUMAR GUPTA

Ph.D., PE, MASCE
Head, Civil Engg. Deptt.
Government Polytechnic,
Washim – 444505

MRS. RUPALI PRATIK KHADTAR

M.E. (Structures)
I/C HOD Department of Civil Engg.
A.I. Abdul Razzak Kalsekar,
Polytechnic, Panvel.

N4515

TRAFFIC ENGINEERING (Civil Engg. Group)   ISBN : 978-93-89533-30-9

First Edition : **September 2019**

© : **Authors**

Published By :

NIRALI PRAKASHAN

Abhyudaya Pragati, 1312, Shivaji Nagar

Off J.M. Road, PUNE – 411005

Tel - (020) 25512336/37/39, Fax - (020) 25511379

Email : niralipune@pragationline.com

➤ DISTRIBUTION CENTRES

PUNE

Nirali Prakashan : 119, Budhwar Peth, Jogeshwari Mandir Lane, Pune 411002, Maharashtra

(For orders within Pune) Tel : (020) 2445 2044, 66022708, Fax : (020) 2445 1538; Mobile : 9657703145

Email : niralilocal@pragationline.com

Nirali Prakashan : S. No. 28/27, Dhayari, Near Asian College Pune 411041

(For orders outside Pune) Tel : (020) 24690204 Fax : (020) 24690316; Mobile : 9657703143

Email : bookorder@pragationline.com

MUMBAI

Nirali Prakashan : 385, S.V.P. Road, Rasdhara Co-op. Hsg. Society Ltd.,

Girgaum, Mumbai 400004, Maharashtra; Mobile : 9320129587

Tel : (022) 2385 6339 / 2386 9976, Fax : (022) 2386 9976

Email : niralimumbai@pragationline.com

➤ DISTRIBUTION BRANCHES

JALGAON

Nirali Prakashan : 34, V. V. Golani Market, Navi Peth, Jalgaon 425001, Maharashtra,

Tel : (0257) 222 0395, Mob : 94234 91860; Email : niralijalgaon@pragationline.com

KOLHAPUR

Nirali Prakashan : New Mahadvar Road, Kedar Plaza, 1st Floor Opp. IDBI Bank, Kolhapur 416 012

Maharashtra. Mob : 9850046155; Email : niralikolhapur@pragationline.com

NAGPUR

Nirali Prakashan : Above Maratha Mandir, Shop No. 3, First Floor,

Rani Jhanshi Square, Sitabuldi, Nagpur 440012, Maharashtra

Tel : (0712) 254 7129; Email : pratibhabookdistributors@gmail.com

DELHI

Nirali Prakashan : 4593/15, Basement, Agarwal Lane, Ansari Road, Daryaganj

Near Times of India Building, New Delhi 110002 Mob : 08505972553

Email : niralidelhi@pragationline.com

BENGALURU

Nirali Prakashan : Maitri Ground Floor, Jaya Apartments, No. 99, 6th Cross, 6th Main,

Malleswaram, Bengaluru 560003, Karnataka; Mob : 9449043034

Email: niralibangalore@pragationline.com

Other Branches : Hyderabad, Chennai

niralipune@pragationline.com | www.pragationline.com

Also find us on www.facebook.com/niralibooks

Preface ...

A text book on '**Traffic Engineering**' is prepared for the benefit of Fifth semester students of Diploma in Civil Engineering Group. The book is written according to the I - Scheme syllabus prescribed by the **Maharashtra State Board of Technical Education.** Though the book is written for Diploma students, it is very informative to Degree, A.M.I.E and Students of autonomous institutes.

Road is important, largest and basic mode of transportation in India. The transportation by roads is the only one mode, which could give maximum service to all. The road is also easy and effective mode of transportation. There is much scope of road development work and its maintenance in our country. The main aim of this book is to make direct approach to the subject, tried to present the topics in simple, informative, readable and lucid form so that even an average students can grasp the subject by self-study. We hope this volume will prove useful to the students and teachers as well as field engineers.

We are thankful to Shri Dineshbhai Furia and Mr. Jignesh. Furia of **Nirali Prakashan** for giving the opportunity to write this book and Mr. Shashikant Patel & Ms. Anagha for carefully reading and kind co-ordination the entire manuscript and offering constructive suggestions. We are indeed thankful to Rahul Thorat (for D.T.P.) and Anjali Mulye (for Fig. Drawing) and entire staff of Nirali Prakashan to bringing out this book.

We express our indebtedness to all books, references, periodicals and journals, which helped in the preparation of this book.

Although every care has been taken to check mistakes, yet it is difficult to claim the perfection. Any errors, omissions and suggestions come forward from teachers and students to improve the subject matter will be thankfully acknowledge and incorporated.

Authors

Syllabus ...

Chapter 1 : Fundamentals of Traffic Engineering
1.1 Traffic Engineering : Definition, Objects, Scope
1.2 Road user's Characteristics-Physical, Mental, Emotional Factors
1.3 Vehicular Characteristics-weight, Length, Height, Speed, Efficiency of Breaks
1.4 Road Characteristics-gradient, Curve of a road, Design Speed, Friction between road and tyre surface
1.5 Reaction time-factors affecting reaction time. PIEV Theory

Chapter 2 : Traffic Studies
2.1 **Traffic Studies :** Types, purpose, Information required for traffic studies.
2.2 **Traffic volume study :** definition, purpose.
2.3 **Methods of collection of traffic volume count data :** manual, automatic recorders, moving car method.
2.4 Necessity of Origin and Destination study and its methods.
2.5 **Traffic volume count data :** Representation and analysis of data.
2.6 **Speed studies :** Spot speed studies, and its presentation.
2.7 Need and method of parking survey.

Chapter 3 : Road Signs and Traffic Markings
3.1 Traffic Control Devices – definition, necessity, types
3.2 Road Signs – definition, objects of road signs.
3.3 Classification as per IRC : 67-1977 – Mandatory of Regulatory, Cautionary or warning, informatory signs, Location of cautionary or warning sign in urban and no-urban areas, Points to be considered while designing the road signs. Points to be considered while erecting the road signs.
3.4 Traffic markings – definition
3.5 Classification of traffic markings – carriage way, kerb, object marking and reflector markers

Chapter 4 : Traffic Signals and Traffic Islands
4.1 **Traffic Signals :** Definition
4.2 **Types of signals :** Traffic control signals, pedestrian signals, special type of traffic signals
4.3 Advantages and disadvantages of traffic control signals
4.4 **Types and traffic control signals :** Fixed time, manually operated, traffic actuated signals
4.5 Location of signals
4.6 Compute signal time by fix time cycle, trail cycle, approximate, Webster's and IRC method and sketch timing diagram for each face.
4.7 **Traffic islands :** definition, advantages and disadvantages of providing islands.
4.8 **Types of traffic islands :** rotary or central, channelizing or refuge
4.9 **Road intersections or junctions :** Definition, Types
4.10 **Intersection at grade :** basic requirements of good intersection at grade, types
4.11 **Grade separated intersection :** advantages and disadvantages, types-over pass or flyovers-Cloverleaf pattern, trumpet type, underpass

Chapter 5 : Road Environment and Arboriculture
5.1 **Street lighting :** Definition, sources necessity, types – luminaire, foot candle, lumen, factors affecting their utilization and maintenance.
5.2 Factors affecting visibility at night.
5.3 **Arboriculture :** definition, objectives, factors affecting selection of type of tree.
5.4 **Maintenance of trees :** protection and care of road side trees

Chapter 6 : Road Accident Studies
6.1 Road Accidents : Definition, types Collision and non-collision accidents
6.2 Causes of accidents
6.3 Measures to prevent road accidents
6.4 Reporting and recording of an accident
6.5 Collision and condition diagram
6.6 Legislation and law enforcement, education and propaganda

Contents ...

☆☆☆

FUNDAMENTALS OF TRAFFIC ENGINEERING

Objectives

1a. Describe the characteristics of road users.

1b. Describe the characteristics for the given type of vehicle.

1c. Calculate reaction time of driver in the given situation.

1d. Explain the factors affecting the reaction time for the given situation.

1.1 TRAFFIC ENGINEERING

Definition of Traffic Engineering as per Institute traffic engineers (USA) :

- The phase of Highway Engineering which deals with planning and geometric design of roads, streets, highway, adjoining lands (abutting lands) and with traffic operations there on for safe, convenient and economic transportation of persons and goods is known as Traffic Engineering.

1.1.1 Objects of Traffic Engineering

(a) To provide free flow of traffic

(b) To provide safety to the traffic

(c) To provide efficient flow of traffic

(d) To provide rapid flow of traffic

(e) To provide economical flow of traffic

(f) To provide convient flow of traffic

1.1.2 Scope of Traffic Engineering

- Traffic Engineering includes study of following phases or areas :
    (a) Traffic characteristics
    (b) Traffic studies
    (c) Traffic analysis
    (d) Geometric design of traffic
    (e) Traffic planning
    (f) Traffic operations
    (g) Traffic regulation
    (h) Traffic control
    (i) Traffic safety
    (j) Traffic administration
    (k) Traffic management

1.1.3 Purpose of Traffic Studies (Role of Traffic Engineer)

(1) To collect the data about the type of traffic and volume of traffic at present.

(2) To calculate or estimate future volume of traffic.

(3) To decide the design of pavement.

(4) To plan and design geometric design of road.

(5) To design drainage system for road.

(6) To design bridges, culverts etc.

(7) To design traffic signals.

(8) To estimate the road taxes.

(9) To manage the traffic administration.

(10) To design the traffic control devices.

1.1.4 Data Required to be Collected during Traffic Studies

(1) Type of traffic

(2) Volume of traffic

(3) Nature of traffic

(4) Origin and destination of traffic

(5) Purpose of traffic

(6) Speed of traffic

(7) Conditions of traffic vehicles

(8) Rate and causes of accidents.

(9) Environmental data

1.2 ROAD USER CHARACTERISTICS

- In traffic, the human is involved as a road user in various ways. Road user is involved as a pedestrian, cyclist, driven of various vehicles and as a motorist.

- So it is important in traffic engineering to study the characteristics and limitations of road users.

  The factors which affects Road user characters are :

  (1) Physical factors

  (2) Mental factors or Psychological factors (emotional)

  (3) Environmental factors

  (1) Physical Factors of Road Users :

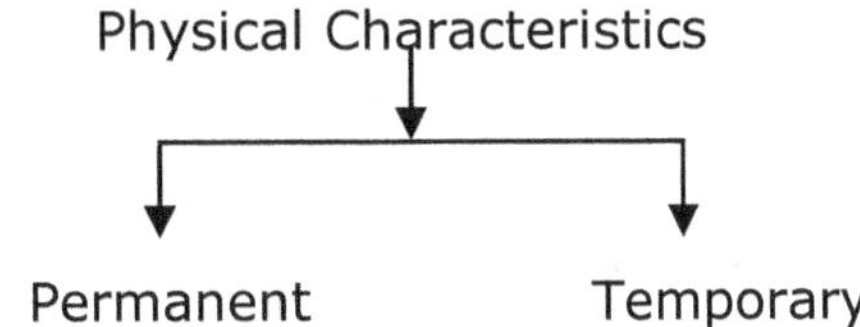

- The permanent characteristics are :

  (i) vision

  (ii) hearing

  (iii)strength

  (iv)general reaction

(1) Vision is one of the important factors that affects almost all aspect of highway design and safety.

- Vision include the acute vision, peripheral vision, eye movement, colom vision, ability to adopt glare, and depth judgment.

- Hearing is an important aid to the road users. It is important for safe driving. It is important for pedestrians and cyclists.

- Strength is useful for parking the vehicle. For maneuvers of heavy vehicle strength is required.

- The general reaction to the traffic situation depends on time required to perceive and understand the traffic situation and after that to take appropriate action. This reactions depends on several factors such as physical condition, psychological condition, speed of vehicle, type of problem etc.

- The temporary physical characteristics are :

  (i)   fatigue

  (ii) alcohol or drugs

  (iii) illness

(2)Mental, psychological and environmental factors : Mental and physiological factors relate to the condition of the mind of the driver; whether he is drunk, fatigued, responsivity of the driver, his anxiousness. These relate to the mind of the driver. To be upright, it should be possible to devise driver worthyness tests just as air worthyness tests for the pilots, so that the wheel of the vehicle is in safe hands.

(3)Environmental factors : These relate to the pattern of traffic flow whether it is mixed, heavy, one-way, hilly etc. and the judgement of the driver to such environment.

1.3 VEHICULAR CHARACTERISTICS

- Road and the vehicle plying on the road could be considered as consenting couple. However a vista of vehicles ply on the road and the static and dynamic characteristics of these vehicles determine certain road elements. For example, the height of the vehicle affects the height of the overbridge.

- The height of driver seat affects the visibility distance and the height of head light determines the head light sight distance at the valley curve. The length of vehicle can affect the capacity and minimum turning radius.

- The dynamic characteristics of the vehicle generally refer to its engine weight, its B.H.P., suspension system etc. These factors to a large extent determine the riding comfort. Keeping these considerations in view IRC has fixed-up maximum dimensions of road vehicles and maximum permissible gross weights. These specifications are summarised in the table below.

Table 1.1: Maximum Dimensions of Road Vehicles

Dimensions of vehicle	Particulars	Maximum dimensions in metres (excluding front and rear bumpers)
Width	All vehicles	2.5
Height	(a) Single decked vehicle for normal application.	3.8
	(b) Double decked vehicle.	4.75
Length	(a) Single unit truck with two or more axles (types 2, 3)	11.0
	(b) Single unit bus with two or more axles (type 2, 3)	12.0
	(c) Semi-trailer tractor combination (Type 2s1, 2s2, 3s1, 3s2)	16.0
	(d) Tractor and trailer combination (Type 2-2, 3-2, 3-3)	18.0

- No combination is allowed to be more than two units and no such combination loaded or otherwise should have overall length exceeding 18 m. The types of road transport vehicles are shown in Fig. 1.1.

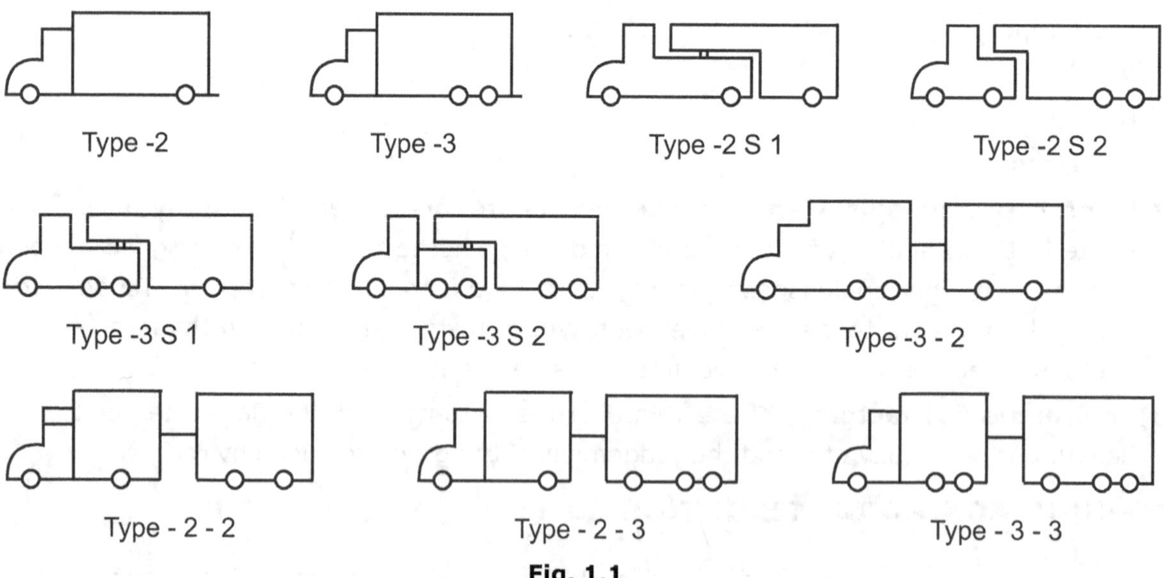

Fig. 1.1

- The weights and other details of the vehicle are as shown in Table 1.2.

Table 1.2: Maximum permissible gross weight and axle weight of transport vehicles

Vehicle type	Maximum gross weight	Maximum axle weight (tones)			
		Truck/Tractor		Trailer	
		FAW	RAW	FAW	RAW
Type 2 (Both Axle single type)	12.0	6	6	–	–
Type 2 (FA single type RA-Dual type)	16.2	6	10.2	–	–
Type 3	24.0	6	18 (TA)	–	–
Type 2 S-1	26.4	6	10.2	–	10.2
Type 2 S-2	34.2	6	10.2	–	18 TA
Type 3 S-1	34.2	6	18 (TA)	–	10.2
Type 3 S-2	42.0	6	18 (TA)	–	18 TA
Type 2-2	36.6	6	10.2	10.2	10.2
Type 3-2	44.4	6	18 (TA)	10.2	10.2
Type 2-3	44.4	6	10.2	10.2	18 TA
Type 3-3	52.2	6	18 (TA)	10.2	18 TA

FAW - Weight on front axle;

RAW - Weight on rear axle;

TA - Tandem axle fitted with 8 tyres

- Braking characteristics of motor vehicles are also very important especially in preventing accidents. Many times a braking test is conducted to access the condition of braking system. To know whether the brakes are OK or not, a vehicle having initial speed of 'u' m/sec. is subjected to full braking for 't' seconds, if the vehicle is brought to stand still position in 'L' metres. Then clearly the braking distance $L = \dfrac{u^2}{2gf}$ on the assumption of the road being level. This equation can thus yield average skid resistance of the pavement surface.

1.3.1 Speed of Vehicle

- The speed of vehicle affects the various geometric design factors of roads. These design factors which affects the speed of vehicle are :

  (a) Design of sight distances

  (b) Design of super-elevation

  (c) Design of length of transition curve

  (d) Design of horizontal curve

  (e) Width of traffic lane

  (f) Design of gradient

  (g) Capacity of traffic lane

  (h) Design of intersections

1.3.2 Efficiency of Tracks

- Efficiency of tracks or tracking characteristics of vehicles depends on the design of tracking and bracking system. The tracking systems are either mechanical or fluid or air tracks.

- The safety of vehicle in a traffic flow is depends on the efficiency of tracking system of vehicle.

- The vehicle operation, stopping distance and specing between the two consecutive vehicles are affected by the efficiency of tracking system of vehicle.

- The acceleration and deceleration of vehicle depends on the efficiency of tracking system of vehicle.

1.4 ROAD CHARACTERISTICS

1.4.1 Design Speed   (S-13, 15, 16, 18; W-13, 16)

- It is the maximum safe speed of a vehicle used for geometrical design of highways. The design speed is permissible for safe and comfortable driving on a given surface of a highway.

- The overall geometrical design of any road depends on design speed. It is essential that the assumed design speed should be in conformity with the high standard of mobility, safety and efficiency desired on different categories of road.

Factors affecting Design Speed

- The value of design speed is governed by the following factors:

    (a) Types of highway,

    (b) Types and condition of the road surface,

    (c) Importance of highway,

    (d) Nature and intensity of traffic,

    (e) Types of curves along the road,

    (f) Sight distance required, and

    (g) Topography of the area etc.

1.4.1.1 IRC Specifications of Design Speed

- The design speed of vehicles on different categories of roads, as recommended by the IRC, is given in Table 1.3.

Table 1.3: Design speed of vehicles on different categories of roads

Sr. No.	Category of road	Design speeds in km/hr							
		Plain terrain		Rolling terrain		Hilly terrain		Steep terrain	
		Ruling	Min.	Ruling	Min.	Ruling	Min.	Ruling	Min.
1.	National and State Highway	100	80	80	65	50	40	40	30
2.	Major District Road	80	65	65	60	40	30	30	20
3.	Other District Road	65	50	50	40	30	25	25	20
4.	Village Road	50	40	40	35	25	20	25	20

- The design speed for low cost roads, which have been recommended on the possibility of further development of road and future expansion of traffic are:

    (a) For hilly area: 32 km/hr, and

    (b) For plain or rolling area: 48 km/hr.

1.4.2 Gradients   (W-09; S-09, 14)

- Gradient is defined as, 'the longitudinal slope (It is the slope of road in longitudinal direction)'. It is the rate of rise (or fall) along the length of road.

- Rising gradient is known as a positive gradient while falling gradient is known as negative gradient.

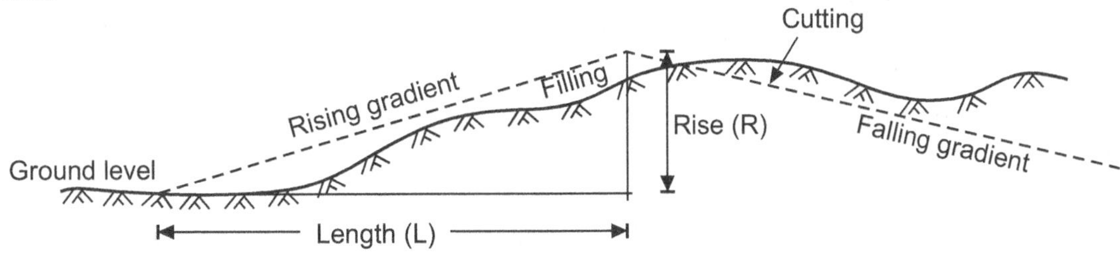

Fig. 1.2

$$\therefore \text{Gradient} = \frac{\text{Vertical distance}}{\text{Horizontal distance}} = \frac{\text{Rise}}{\text{Length}} = \frac{R}{L} \times 100\%$$

- The gradient is expressed as a ratio of length in which the elevation of the centre of road rises or falls by one unit for example, $1 = 100$. It is also expressed as a percentage; thus $1:100$ gradients are expressed as one per cent gradient.

- Gradients on the road are necessary due to the nature of terrain through which the road passes. While travelling along a rising gradient, extra effort or extra fuel consumption is required. Steep gradient should therefore be avoided.

- The gradient to a road is necessary to drain-off the rainwater easily through the side drains. Even in plain areas, the flat gradients of 1 in 250 to 1 in 500 are provided for satisfactory drainage and to keep the road in good condition with less maintenance cost.

1.4.2.1 Purpose of Providing Gradient/Significance of Road Gradient   (W-08, 13, 16)

- The gradients to a road are provided for the following purposes:
  (a) To connect two roads, which are at different levels.
  (b) To provide effective drainage of rainwater falling over the road surface.
  (c) To make the earthwork of the road project economical since a perfectly levelled road involves more cutting and filling.
  (d) To reduce the maintenance cost of the road surface.
  (e) To construct side drains economically with convenient depth below the ground level.

1.4.2.2 Factors Affecting the Gradient   (W-08, S-14)

- Following are the various factors that affect the selection of gradient in the alignment of road:
  (a) The nature of traffic,
  (b) The nature of ground,
  (c) The road-railway intersections and bridge approaches,
  (d) The type of road surface,
  (e) The drainage required,
  (f) The total height to be covered,
  (g) The safety required.

1.4.2.3 Types of Gradients

- Road gradients can be categorized as below:

  (a) Ruling gradient: This is the gradient, which should not be normally exceeded. Vehicles find it convenient and economical to travel along gradient not exceeding this limit. Road designer should therefore aim at not exceeding this limit. The limit is also called design gradient, of course wherever possible lesser gradients should be provided.

Factors governing ruling gradient;

(i) The nature of vehicle.

(ii) The type of road surface.

(iii) The local topographical conditions.

(b) Limiting gradient: This gradient is provided where the topography of the area makes ruling gradients very costly. (S-13)

(c) Exceptional gradient: As the name suggests this gradient is provided only under exceptional circumstances and that too for a short length of the road.

(d) Floating gradient: If a vehicle is along a ascending gradient at constant speed and comes across a descending gradient such that it maintains the same speed without any tractive effort or without any application of breaks, then such a gradient would be known as a floating gradient.

1.4.2.4 IRC Specifications for Different Gradients

- The following table indicates the recommendations in respects of different gradients.

Table 1.4: Recommendations for Different Gradients

Sr. No.	Nature of terrain	Ruling gradient	Limiting gradient	Exceptional gradient
1.	Plain	1:30	1:20	1:15 (For a stretch not exceeding 100 m)
2.	Rolling	1:30	1:20	1:15 (For a stretch not exceeding 100 m)
3.	Mountains	1:20	1:67	1:14.3
4.	Steep	1:20 to 1:16.7	1:16.7 to 1:14.3	1:14.3 to 1:12.5

- A minimum longitudinal gradient is required to be provided for facilitating drainage. The recommended values are given in the Table 1.5.

Table 1.5: Minimum Gradients for Drainage

No.	Particulars	Gradient
1.	Longitudinal drainage along the curved edge of road.	(Minimum) 1:200
2.	Drainage along the road side ditch.	1:100

1.4.3 Friction Between Road and Tyre

- When the vehicle tyers roll over the road surface, the irregularities, uneven surface and roughness of the surface develop the friction between road and tyre.
- This friction is called as rolling resistance. Shocks and impacts are also caused by motion and it affects the motion of vehicle.
- The rolling resistance is depends upon type of surface.
- The rolling resistance (friction) is given by,

$$P_f = mfg$$

P_f = Rolling resistance (N)

m = mass of vehicle (kg)

f = coefficient of rolling resistance

g = acceleration due to gravity (m/sec^2)

- The rolling resistance also depends on the speed of vehicles.

- Upto 50 kMPH value of Rolling resistance is constant.

Table 1.6 : Values of Rolling resistance

Type of Road Surface	Rolling Resistance
(A) Cement concrete	0.01
(B) WBM (good condition)	0.025
(C) WBM (bad condition)	0.037
(D) Grave	0.046
(E) Earth	0.055

1.5 REACTION TIME

- Reaction Time of the driver is the time taken from the instant (time) the object is visible to the driver to the instant (time) the breaks are effectively applied by the driver.
- Reaction time of the driver depends on so many factors.
- The total reaction time is the addition of (i) Perception Time, (ii) Brake Reaction Time

    (i) Perception Time : The perception time is the time required for a driver to come to the realization that the brakes must be applied. The perception time varies from driven to driver. It also depends on distance of objects and environmental conditions.

    (ii) The Brake Reaction Time : The brake reaction time is that time lag between the perception of danger and the effective application of brakes.

    This time is depends on skill of the driver, type of problem and environmental factors.

    Total Reaction Time (t) = Perception Time + Brake Reaction Time

    As per AASHTO policy and Indian Road Congress (IRC), the Total Reaction time is 2.5 seconds.

    The Total Reaction time is explain in PIVE theory.

- The total reaction time of an average driven may vary from 0.5 second for time situation and more than 3 to 4 seconds or even more for complex situations.

1.6 PIEV THEORY

- As per PIEV theory, the total reaction time of the driver is divided into four parts :

    (i) Perception (P)   (ii) Intellection (I)

    (iii) Emotion (E)   (iv) Volition (V)

    (i) Perception (P) : Perception Time (P) is the time required to receive the sence by eyes or ears of a driver towards the brain through the nervous system and spinal cord.

    It is the time required to perceive of an object or situation by the driver through sence.

    (ii) Intellection Time (I) : Intellection Time (I) is the time required for a driver to understand the situation occurs while driving.

    (iii) Emotion Time (E) : It is the time elapsed during fear, anger or any other emotional feelings with reference to the situation.

    This time varies from driver-to-driver.

    (iv) Volition Time (V) : It is the time taken by driver for the final action.

    Final action may be the application of brake or any other avoiding action like turning etc.

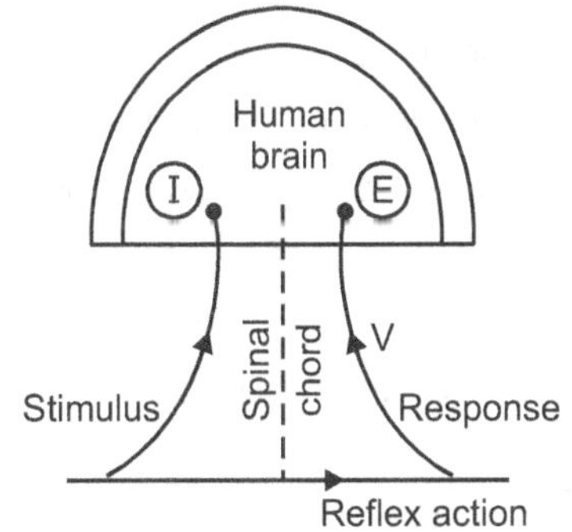

Fig. 1.3 : Reaction Time and PIEV Process

Important Points

- **Traffic Engineering as per Institute traffic engineers (USA) :** The phase of Highway Engineering which deals with planning and geometric design of roads, streets, highway, adjoining lands (abutting lands) and with traffic operations there on for safe, convenient and economic transportation of persons and goods is known as Traffic Engineering.

- Road user is involved as a pedestrian, cyclist, driven of various vehicles and as a motorist.

- The permanent characteristics are :

  (i) vision   (ii) hearing

  (iii) strength   (iv) general reaction

- Vision is one of the important factors that affects almost all aspect of highway design and safety.

- Efficiency of tracks or tracking characteristics of vehicles depends on the design of tracking and bracking system. The tracking systems are either mechanical or fluid or air tracks.

- Design Speed : It is the maximum safe speed of a vehicle used for geometrical design of highways. The design speed is permissible for safe and comfortable driving on a given surface of a highway.

- The gradient is expressed as a ratio of length in which the elevation of the centre of road rises or falls by one unit for example, 1 = 100. It is also expressed as a percentage; thus 1:100 gradients are expressed as one per cent gradient.

- Reaction Time of the driver is the time taken from the instant (time) the object is visible to the driver to the instant (time) the breaks are effectively applied by the driver.

- Volition Time (V) : It is the time taken by driver for the final action.

- When the vehicle tyers roll over the road surface, the inequalities, uneven surface and roughness of the surface develop the friction between road and tyre.

Practice Questions

1. Write short note on Traffic Engineering.
2. State the objects of Traffic Engineering.
3. Explain : Role of Traffic Engineering OR State the purpose of Traffic Engineering.
4. Write short note on scope of Traffic Engineering.
5. State the factors which affects the speed of vehicle.
6. Explain Efficiency of tracks.
7. State the factors affecting Design Speed.
8. Define :

   (i) Gradient   (ii) Reaction Time   (iii) Perception Time

   (iv) Emotion Time   (v) Intellection Time

9. State and explain types of gradients.
10. Explain : Friction between Road and Tyre.
11. Enlist Road Users characteristics.
12. Enlist vehicular characteristics.
13. Enlist different types of curve. Explain any one.
14. Give function value of any four Road Surface.
15. Explain PIEV theory with sketch.
16. Define design speed. Give its IRC specifications.

☆☆☆

Chapter **2**...

TRAFFIC STUDIES

Objectives

2a Conduct traffic volume studies for the given section of road.

2b Analyze origin-destination studies data collected for the given road.

2c Analyze spot speed study data collected for the given road.

2d Design and develop the parking system for the given situation.

2e Suggest the improvement in road geometries bases on traffic volume count.

2.1 TRAFFIC STUDIES

Fig. 2.1

2.1.1 Definition

- Traffic study is a planning tool/method used by transportation/Traffic engineers to predict demands on the transportation network and subsequently to determine transportation improvements that may be necessary to adjust/accommodate new development or to find a solution to a new traffic issue/transportation problem. Thus a traffic study is a detailed examination and analysis of a transportation system supported by data collection to examine traffic characteristics for safe and efficient traffic movement.

- The process is as follows:
  - o A study starts with the identification and defining/scoping of a transportation problem, followed by data collection and analysis.
  - o A study is typically performed to explore a specific aspect of, or question about, a transportation system.
  - o Study results are usually summarized in a report.

Types of Traffic Studies :

- There are different types of studies which are required to be carried out for specific purposes and are as follows:
  - o Traffic volume study
  - o Speed and delay study
  - o Origin and destination study
  - o Traffic flow characteristics study
  - o Traffic capacity and level of service study
  - o Parking study
  - o Accident study

2.1.2 Purpose

- Transportation studies serve :
  - o To quantify (giving a numerical value) the extent of a transportation or traffic problem.
  - o And/or to provide/check an analysis of a proposed transportation solution whether the solution suffices or not.

2.1.3 Inforamtion Required for Traffic Studies

- The following are some conditions that may necessitate conducting a traffic study:
  - o When a new development is proposed and it will generate substantial new traffic;
  - o When financial recoveries/assessments are implemented (toll fee etc.);
  - o When a major roadway improvement or reconstruction project is proposed;
  - o When existing transportation problems are evident, such as a high crash location or at a location with complex roadway geometrics; extraordinary delays at a junction
  - o When a development is proposed for a sensitive area with regard to environment or political; and
  - o At the judgment or discretion of jurisdiction staff/transportation officials based on unusual circumstances.

2.2 TRAFFIC VOLUME/COUNT STUDY

- Traffic volume studies are conducted to determine the number, movements, and classifications of roadway vehicles at a given location. These data can help identify critical flow time periods, determine the influence of large vehicles or pedestrians on vehicular traffic flow, or document traffic volume trends. The length of the sampling period depends on the type of count being taken and the intended use of the data recorded.

- For example, an intersection count may be conducted during the peak flow period. If so, manual count with 15-minute intervals could be used to obtain the traffic volume data
- A **traffic count** is a count of vehicular or pedestrian traffic, which is conducted along a particular road, path, or intersection. A traffic count is commonly undertaken either automatically (with the installation of a temporary or permanent electronic traffic recording device), or manually by observers who visually count and record traffic on a hand-held electronic device or tally sheet. Traffic counts can be used by local councils to identify which routes are used most, and to either improve that road or provide an alternative if there is an excessive amount of traffic. Counting is the most fundamental measurement in traffic engineering. Counting of vehicles, counting of persons etcetera are required for various reasons. It may be person or it may be vehicle but counting is the elemental requirement as to how many people, how many cars or what would be the vehicular traffic in general. Hence, counting techniques are used to produce estimates of traffic volume, rate of flow, demand and capacity. The term traffic volume study can be termed as traffic flow survey or simply the traffic survey. It is defined as the procedure to determine mainly volume of traffic moving on the roads at a particular section during a particular time.
- It may be vehicle or person passing a point during a specified time period which is usually one hour but may not be always the case. It means during a period of time which can be fifteen minutes, can be half an hour or even sixty minutes or one hour and how many vehicles or persons are passing a point during that particular time period is known as volume.

2.2.1 Definition of Volume Study

- Traffic volume study is the quantity of vehicles crossing a section of road per unit time at any selected period. It is usually expressed in terms of Passenger Car Unit (PCU) and measured to calculate Level of Service of the road and related attributes like congestion, carrying capacity, V/C Ratio, identification of peak hour or extended peak hour etc.
- There are two important component of finding the traffic volume, one is the number of vehicle and the second is the unit time. In traffic volume studies we also measure the demand actually. It is number of vehicles or persons who desire to travel past a point. Volume studies are carried out primarily at mid block locations and at intersections. These are the two places where normally carry out traffic volume studies.
- Next point is what should be period of volume studies, for how much period we must carry out the traffic volume studies. This again depends on the purpose and use of the data. What is the reason or why we are carrying out traffic volume studies? It may be like peak hour or peak period, morning peak, evening peak or may be twelve hour volume count, it may be sixteen hour volume count, it may be twenty four hour volume count, or may be seven day volume count. Hence it is possible to have different traffic volumes.

2.2.2 Purpose of Volume Study

- The purposes/uses of traffic studies are as follows:
    - Planning (Establishing the use of the road network by vehicles of different categories, traffic distribution, PCU/vehicle value, Need of median shifting or road widening etc.).

o   Traffic operation and control (Checking the efficiency/saturation of the road network by comparing current traffic volume with the calculated capacity or by identifying level of service).

o   Traffic pattern (draw inferences on the basis of data collected).

o   Structural design of pavement (provide possible solutions and improvement suggestion for the problem identified).

o   Regulatory measures (identifying the hourly distribution of vehicles and peak hour, identify level of service and compare modal composition on different hierarchy of roads).

2.2.3 Duration and Interval of Traffic Counts

- In order to predict traffic flow volumes that can be expected on the road network during specific periods, knowledge of the fact is required that traffic volumes changes considerably at each point in time. There are three important cyclical variations:

  1. **Hourly pattern:** the way traffic flow characteristic varies throughout the day and night;

  2. **Daily Pattern:** The day-to-day variation throughout the week.

  3. **Monthly and yearly Pattern:** The season-to-season variation throughout the year.

     When analysing the traffic one must also be aware of the directional distribution of traffic and the manner in which its composition varies as it is important to deal with tidal flow.

  4. **Hourly patterns :** Typical hourly patterns of traffic flow, particularly in urban areas, generally show a number of distinguishable peaks. Peak in the morning followed by a lean flow until another peak in the middle of the afternoon, after which there may be a new peak in the late evening. The peak in the morning is often more sharp by reaching the peak over a short duration and immediately dropping to its lowest point. The afternoon peak on the other hand is characterised by a generally wider peak. The peak is reached and dispersed over a longer period than the morning peak.

  5. **Daily patterns :** The traffic volume generally varies throughout the week. The traffic during the working days (Monday to Friday) may not vary substantially, but the traffic volume during the weekend is likely to differ from the working days on different type of roads and in different directions

2.3 TRAFFIC VOLUME COUNT DATA

2.1.3 Methods of Collection of Traffic Volume Count Data

- Number of vehicles (traffic volume) can be counted either manually or by automatic machines depending upon various factors like duration of count, accuracy required, location of traffic study area, manpower available, budget, technology/instrument available, magnitude of traffic data required or to be collected etc.

2.3.1.1 Manually

- The manually method is the method in which a group of peoples are trained to record the total numbers of vehicles crossing a section of the road in a specified period of time on record sheets manually. It is the simplest form in which an observer counts the numbers of vehicles along with its type, passing through a section for a definite time interval. It is normally used for light traffic volumes. For light volumes, tally marks on a printed form are sufficient. Raw data from those inventories is then organized for compilation and analysis. This method of data collection can be expensive as manpower is used in this method, but it is otherwise necessary in most cases where vehicles are to be classified with a number of movements recorded separately, such as at intersections.

ROADS AND HIGHWAYS DEPARTEMENT   **TRAFFIC COUNT TALLY SHEET**   sheet : ...of...

Name of Road :.............................. Road No.: .... Duration From :.......................... To :.............................

Station Name :............................... Station Number :................................   Date : ....../......../............
                                                                                                      DD   MM   YY

Enumerator :.................................... Supervisor :..

Hours counted	MOTORISED										NON-MOTORISED		
	1	2	3	4	5	6	7	8	9	10	11	12	13
	Heavy truck	Medium truck	Small truck	Large Bus	Mini bus	Microbus	Utility	Car	Auto Rickshow	Motor cycle	Bicycle	Cycle Rickshow	Assist/Park cars
: to :													
: to :													
: to :													

Fig. 2.2

- Advantage of manual method is that it gives the full detail of the traffic like various classes, stream and turning movement etc of the vehicles and also it is good to take some manual observations for every counting for checking the instruments accuracy. Disadvantage is that it cannot work throughout the day and night for all days of year; it is labour intensive, therefore costly and prone to human error.

2.3.1.2 Automatic Count

- This method is employed in cases where manual count method is not feasible. These can be used to obtain vehicular counts at non-intersection points. Total volume, directional volume or lane volumes can be obtained depending upon the equipment used. Various instruments are available for automatic count, which have their own merits and demerits. Some of the widely used instruments are pneumatic tubes, inductive loops, weigh-in-motion Sensor, micro-millimeter wave Radar detectors and video camera. To permanently or temporarily monitor the usage of a road, an electronic traffic counter can be installed or placed to measure road usage continuously or for a short period of time. Most modern equipment called ATR's (Automatic Traffic Recorders) store count and/or classification data recorded in memory in a time duration or interval fashion that can be downloaded and viewed in software or via a count display on some equipment. Some of the widely used automatic instruments are :

(i) Pneumatic tubes : These are tubes placed on the top of road surfaces at locations where traffic counting is required. As vehicles pass over the tube, the resulting compression sends a burst of air to an air switch.

Fig. 2.3 : Pneumatic Road Tube Counter with the Tube on The Roadway and the Counting Device on The Sidewalk

(ii) Inductive loops : Inductive loop detector consists of embedded turned wire. It includes an oscillator, and a cable, which allows signals to pass from the loop to the traffic counting device. Inductive loops are cheap, almost maintenance-free and are currently the most widely used equipment for vehicle counting and detection.

Fig. 2.4 : Traffic counter system using inductive loops connected to a cabinet with solar panels and 3g modem to transmit traffic information

(iii) Weigh-in-Motion Sensor types : A variety of traffic sensors and loops are used to count, weigh and classify vehicles while in motion, and these are collectively known as Weigh In Motion (WIM) sensor systems. Some notable traffic sensors are:

- **Bending Plates** which contains strain gauges that weigh the axles of passing vehicles

- **Capacitive Strip** is a thin and long extruded metal used to detect passing axles. Capacitive strips can be used for both statistical data and axle configuration.

- **Capacitive Mat** functions in a similar manner as the capacitive strip but it is designed to be mobile and used on a temporary basis only.

- **Piezo-electric Cable** is a sensing strip of a metallic cable that responds to vertical loading from vehicle wheels passing over it by producing a corresponding voltage. The cable is very good for speed measurement and axle-space registration, and is relatively cheap and maintenance.

(iv) Micro-millimeter wave Radar detectors : Radar detectors actively emits radioactive signals at frequencies ranging from the ultra-high frequencies (UHF) of 100 MHz, to 100 GHz, and can register vehicular presence and speed and can be used determine vehicular volumes and classifications in both traffic directions.

Fig. 2.5 : A Radar-Based Traffic Counter Powered By a Solar Panel

(v) Video Camera : Video image processing system utilize machine vision technology to detect vehicles and capture details about individual vehicles when necessary. The system is useful for traffic counting and give a +/- 3% tolerance, and is not appropriate for vehicular speed and their classification.

Fig. 2.6 : Bike Counter With Display Showing the Number of Bikes

2.3.1.3 Moving Car/Vehicle/Observer Method

- Moving car or moving observer method of traffic stream measurement has been developed to provide simultaneous measurement of traffic stream variables. This method is the most commonly used method to get the relationship between the fundamental stream characteristics. In this method, the observer moves in the traffic stream unlike all other previous methods.

- In this method, the observer moves in the traffic stream and makes a round trip on a test section. The observer starts at section, drives the car in a particular direction say northward to another section, turns the vehicle around drives in the opposite direction say southward toward the previous section again. Let, the time in minutes it takes to travel north (from X-X to Y-Y) is t_n, the time in minutes it takes to travel south is t_s, the number of vehicles travelling north in the opposite lane while the test car is travelling south be m_n, the number of vehicles that overtake the test car while it is travelling south be m_o, and the number of vehicles that the test car passes while it is traveling south from be m_p.

- The volume (q_s) in the southbound direction can then be obtained from the expression and

$$q_s = (m_n + m_o - m_p)/(t_s + t_n)$$

the average travel time in the southbound direction is obtained from

$$t_{s\,(avg)} = t_s - (m_o - m_p)/q_s$$

One can find the three fundamental stream parameters viz. speed, flow and density with this method and can find the relationships among them.

2.4 TRAFFIC VOLUME COUNT DATA – REPRESENTATION AND ANALYSIS OF DATA

- Traffic volume count/traffic flow data are first measured and then converted into an equivalent vehicle unit known as PCU which depends upon the highway capacity used by a vehicle individually or the impact created by a vehicle to consume the portion of a highway.

2.4.1 Passenger Car Unit (PCU)

- Passenger Car Unit (PCU) is a unit used in Transportation Engineering, to assess traffic-flow rate on a highway and also used for expressing highway capacity. A Passenger Car Unit is a measure of the impact that a mode of transport has on traffic variables (such as headway, speed, density) compared to a single standard passenger car. This is also known as passenger car equivalent. For example, typical values of PCU (or PCE) are:
  - o private car (including taxis or pick-up) = 1
  - o motorcycle = 0.5
  - o bicycle = .2
  - o horse-drawn vehicle = 4
  - o bus, tractor, truck = 3.5
  - o Bullock Cart = 6
  - o Bullock cart Large = 8
  - o HCV = 3.5
  - o LCV = 2.2

Highway capacity is measured in PCE/hour daily

- A common method used in the US is the density method. However, the PCU values derived from the density method are based on underlying homogeneous traffic concepts such as strict lane discipline, car following and a vehicle fleet that does not vary greatly in width.
- On the other hand, highways in India, carry heterogeneous traffic, where road space is shared among many traffic modes with different physical dimensions. Loose lane discipline prevails; car following is not the norm. This complicates computing of PCE.
- Using multiple heuristic techniques, transportation engineer converts a mixed traffic stream into a hypothetical passenger-car stream.

2.4.2 Representation and Analysis of Data

- The flow data collected by a traffic/transportation engineer observed on field can be classified into three types depending upon the length of observation:
  - o Measurement at a point
  - o Measurement over a short section

- o　Measurement over a long section
- Out of these, flow is the main traffic parameter measured at a point. Flow can be defined as the number of vehicles passing a section per unit time. Traffic volume studies are mainly carried out to obtain actual data concerning the movement of vehicles at selected point on the street or highway system.

Types of Volume Measurement

- Volume count varies considerably with time. Hence, several types of measurement of volume are commonly represented to average these variations. There are so many ways in which these traffic volume data may be represented. These are as described below:

(a) Average Annual Daily Traffic (AADT) : This is given by the total number of vehicles passing through a section in a year divided by 365. This can be used for following purposes:
- o　Measuring the present demand for service by the street or highway.
- o　Developing the major or arterial street.
- o　Evaluating the present traffic flow with respect to the street system.
- o　Locating areas where new facilities or improvements to existing facilities are needed.

(b) Average Annual Weekday Traffic (AAWT) : This is defined as the average 24-hour traffic volume occurring on weekdays over a full year.

(c) Average Daily Traffic (ADT) : An average 24-hour traffic volume at a given location for some period of time less than a year. It may be measured for six months, a season, a month, a week, or as little as two days. An ADT is a valid number only for the period over which it was measured.

(d) Average Weekday Traffic (AWT) : An average 24-hour traffic volume occurring on weekdays for some period of time less than one year, such as for a month or a season.

(e) Traffic composition : There are two broad classes; fast moving vehicle and slow moving vehicle. Under fast moving vehicle there are other different vehicle classifications like two wheelers, three wheelers and four wheelers etc. Slow moving consists of carts, bicycles etc.

(f) Design Hourly Volume : It is the economic hourly flow of future year, which is used for designing geometric features of roadway. It is chosen in such a way that during the design period it should not be exceeded too often or too much.

(g) Rate of Flow : Rate of flow is used to express an equivalent hourly rate for vehicles passing a point along a roadway or for traffic during an interval less than 1-hr (usually 15 min)

(h) Saturation flow : The maximum hourly rate of an approach at a signalized junction.

(i) Service flow rate : The maximum hourly rate of traffic flow of a roadway section during a given period under prevailing roadway condition.

(j) Forced flow : When lane changing opportunity decreases with increasing traffic volume and drivers are forced to follow slow leaders.

(k) Free flow : When drivers face no restriction in driving and can maintain their desired speeds.

(l) Peak flow : Flow at peak periods.

(m) Off-peak flow : Flow at off-peak/lean periods.

(n) Stable/ Steady flow : When demands are well below the roadway capacity and the average rate of flow remains almost constant with time.

(o) Unstable flow : When demand is at or near or exceeds the roadway capacity and the average rate of flow fluctuates largely with time.

(p) Contra flow : For repair works; an arrangement on a large road by which traffic going in both directions uses only one side of the road.

(q) For bus priority : a special arrangement on one-way street by which only bus is allowed to go in opposite direction.

(r) Tidal flow : When traffic flows in both direction exhibit unbalanced characteristics at peak periods, for example, morning rush at in-bound lanes due to commuter traffic and in the evening the same is true for the out-bound lanes.

(s) Peak Hour Volume : the traffic volume measured over the busiest hour of the day at a given location. PHV gives the highest hourly volume in a day

(t) Peak Hour Factor : It is the average volume during the peak 60 minute period divided by four times the average volume during the peak 15 minute's period.

$$PHF = \frac{V_{av}^{60}}{4 \times V_{av}^{15}}$$

$$PHF = \frac{(average\ flow\ rate)}{(4 * Peak\ 15\ minute\ flow\ rate)}$$

- Following is an example of how the peak hour factor is computed and how it might affect the final results of a capacity calculation.

- The table below shows flow rates that were measured for four 15-minute time periods for each of the 12 intersection movements. Examination of this table shows that second time period, which begins at 4:15 pm, is the peak 15-minute period of the four that are shown here. The total flow for this time period is 4,220 veh/15 minutes, or 16,880 veh/hr. The average flow rate for the hour is 12,640 veh/hr; this is the sum of the total volumes observed during each of the four 15-minute periods shown below. The peak hour factor can then be computed as follows:

$$PHF = (average\ flow\ rate)/(4*Peak\ 15\ minute\ flow\ rate)$$
$$= 12,640/16,880 = 0.75$$

Time period	Eastbound			Westbound			Northbound			Southbound			Total
	LT	TH	RT	LT	TH	RT	Lt	TH	RT	LT	TH	RT	
4:00pm	40	55	175	50	50	75	120	815	45	40	700	55	2,220
4:15 pm	50	75	375	55	80	125	215	1,025	20	60	1,975	165	4,220
4:30 pm	30	75	125	45	75	115	20	975	35	55	1,200	145	2,895
4:45 pm	45	60	175	55	85	150	145	1,015	45	50	1,350	130	3,305

- You can see that the possible values of PHF can range between 0.25 and 1.00, inclusive. Higher numbers indicate a flatter peak. It is rare that PHF drops much below a value of about 0.70. In this case, the PHF of 0.75 is indicative of a very sharp peak for an urban environment, and is probably more characteristics of small towns and cities than larger urban areas.

(u) Design Hour Volume (DHV): Peak hourly volumes are different for every day of the year. The thirtieth highest peak hour volume is considered for rural design and the fiftieth highest peak hour volume is considered for urban design and are often called DHV. The DHV is a two-way traffic volume that is determined by multiplying the ADT by a percentage called the K-factor. Values for K typically range from 8 to 12% for urban facilities and 12 to 18% for rural facilities.

$$DHV = AADT \times k$$

where, k denotes the proportion of daily volume occurring during the peak hour (expressed as decimal) k is often the ratio of 30th or 50th HV to the AADT from a similar site.

2.5 NECESSITY OF ORIGIN AND DESTINATION STUDY AND ITS METHODS

2.5.1 Necessity

- In a transportation study, it is often necessary to know the exact origin and destination of the trips. The information yielded by O-D survey includes land-use of the zones of origin and destination, household characteristics of the trip making family, time of the day when journeys are made, trip purpose and mode of travel.

- Following are the objects/necessities/Use:
  1. Origin-destination (O-D) surveys provide a detailed picture of the trip patterns and travel choices of a city's or region's residents.
  2. These surveys collect valuable data related to households, individuals and trips.
  3. This information allows stakeholders to understand travel patterns and characteristics and to understand trends.
  4. To provide input to travel demand model development, forecasting, and planning for area-wide transportation infrastructure needs and services.
  5. To monitor progress in implementing transportation policies.
  6. To establish preferential routes for various categories of vehicles
  7. To find location of new proposed road
  8. To locate expressway
  9. To regulate movement of heavy vehicles
  10. To locate new bridge as per traffic demands

2.5.2 Methods of O and D Study

- The most important factor for a successful survey is the level of participation by public upon which the success of any method depends. Therefore, reliable assistance of as many respondents as possible is the key to a successful survey. Some of thing which can be done for a respondent friendly survey are:
  - Design the questionnaires in a type size that people can read, a clear layout, and understandable questions.
  - Keep the questionnaires as short as possible
- Various Origin and Destination survey methods are described as follows:

(A) Roadside Interview:

Fig. 2.7

- In this method interview stations are decided beforehand and the vehicles are stopped at the interview station by the surveyors and questionnaire is supplied to the drivers and answers are collected. Hence this interview includes directing vehicles into a designated area and asking a series of short questions. This technique is widely used and has a very high response rate but sometimes implementation is difficult due to disruption of traffic. In this method investment cost is low but requires high labour and personnel requirements. This method is quick but vehicles are stopped for interview and there is obstruction in the traffic flow. This method has a broad geographical coverage as it includes vehicles from outside the study area also, but implementation can be for limited locations and hence sampling may be biased.

- Information to be collected in this survey is as follows:
  - Place and time of origin
  - Place and time of destination
  - Route
  - Purpose of the trip
  - Types of vehicles
  - Number of passengers in each vehicle

Survey Questionnaire (Standard format for O & D Survey)

- Please take a moment to answer a few questions about your trip. Your responses will help determine the need for improvements in this area.

1. Where did your trip begin?

   City/Town------State   Pin

2. What type of place is your trip start point?

   ❐ Primary Residence ❐ Workplace ❐ Store ❐ School (I am a student) ❐ Recreation Area

   ❐ Other ____

3. Where did your trip end?

   City/Town ______________________________ State _______ Zip __________

4. What type of place is your trip end point?

   ❐ Primary Residence ❐ Workplace ❐ Store ❐ School (I am a student) ❐ Recreation Area

   ❐ Other

5. What was the purpose of your trip?

   ❐ Work Commute ❐ Business Related ❐ Shopping ❐ School (attend class) ❐ Recreation

   ❐ Other

6. What type of vehicle were you in?

   ❐ Passenger vehicle (car, motorcycle, SUV, pick-up truck, minivan) ❐ Commercial vehicle

   ❐ Other ___

7. Do you regularly use this route? ❐ Yes ❐ No

   Please add any comments on transportation you May have.

   Comments ______________________________

Thank you very much for your co-operation!

(B) License plate Mail-out surveys :

- The license plate mail-out survey involves recording license plate numbers of vehicles on a selected roadway, tracing vehicle ownership, and mailing a survey to owners. There are different methods for tracing the license plate number: taking a photo/video or manually recording the tag on vehicles. Photo/video are often used for high volume highways. Labour requirements are more for tracing the ownership of vehicles and also may be less accurate but no disruption of traffic occurs.

(C) Telephone survey :

- In this type of survey, appropriate vehicle owners are contacted on phone and are interviewed through a series of short questions. No disruption of traffic occurs but has high personnel requirements as compared to mail surveys. Some other advantages are low investments and saving the time This method has many disadvantages also like: appropriate sampling is difficult, low responses may create biased data, information responses not good, there is poor coverage of vehicles licensed in other states and areas. Now-a-days in this method, interview are conducted with Computer Assisted Telephone Interview Technology (CATI) A complete CATI system includes automatic dialing of next household to the interviewer to ask the nest question automatic skipping and branching within the list of questions depending on the answer to the previous question, immediate logic checks on answer provided.

(D) Taxi survey :

- Large urban areas usually have a sizeable amount of travel by taxis. In such cases, a separate taxi survey is necessary. The survey consists of issuing questionnaires or log sheets to the taxi drivers and requesting them to complete the same.

(E) Post card/ Mail surveys :

- In this method reply-paid questionnaires are handed over to each of the drivers at the survey points and requesting them to complete the information and return by post. This method is simpler and cheaper than many others. Disadvantage is that response may not be good. In this type of survey there is no disruption of traffic and also low investments are required. But obtaining trip details is difficult and also response rates are low which may create a biased data. Sampling is also difficult.

(F) Online survey :

- Web-based surveys as compared to other surveys provide more valuable information less expensively. Response rate of these surveys can be very high if proper incentives are provided, as the money saved for data input and validation can be used to boost up the response. Out of all Origin Destination survey methods these surveys provide ideal solution for information gathering because of their fast turnaround and can cover very large sample size. Sampling is difficult in this method though the response rate is very high and no disruption of traffic takes place.

(G) Home-interview survey :

- In this method random sample of 0.5 to 10% of the population is selected and the residences are visited by the trained persons who collect the travel data from each member of the household. The detailed information regarding the trips made by the members is obtained on the spot. The data collected will be useful for planning the road network and other roadway facilities.
- Data collection:
  - number of trip made
  - their origin & destination
  - purpose of trip

- o　travel mode
- o　number of residents
- o　age
- o　vehicle ownership
- o　number of drivers
- o　family income

- Advantages include the problem of stopping of vehicles and consequent difficulties are avoided. The present travel needs are clearly known and the analysis is also simple. Additional data including socio-economic and other details may be collected so as to be useful for forecasting traffic and transportation growth.

(I) GPS Receiver :

- Cell phone tracking provides data on phone (owner's) movements as cell phone transitions from one cell tower to another. But widespread utilization of GPS receivers for O-D data collection is currently cost prohibitive, especially for large rural and urban areas. Though there is no disruption of traffic, it requires very high equipment cost. This method is limited to samples only in the study area and also information regarding trip purpose is limited.

(J) Tag survey :

- In this method at each point where the roads cross the cordon/boundary line, vehicles are stopped and a tag is affixed, usually under a wind screen wiper. The tags for different surveys stations have different shapes/colour to identify the survey station. The vehicles are stopped again at the exit points where the tags are removed. The time of entering and leaving the area may be marked on the tags in order to enable the journey time to be determined. This method is simple and errors are not very much. It is not possible to handle all the vehicles; sampling may have to be done.

(K) Public Transport Surveys :

- In this method interviewer may enter the vehicle and carry out the interviews when the vehicles is in motion. Post-card questionnaires eliminate delays, but are likely to evoke poor response or contain an element or bias. These questionnaires may also be collected at the stations inside the survey area.

(L) Commercial Vehicle Surveys :

- Commercial vehicle surveys are conducted to obtain information on journeys made by all commercial vehicles based within the study area. The addresses of the vehicles operators are obtained and they are contacted. Survey forms are issued to drivers with a request that they record particulars of all the trips they would make.

2.6　SPEED STUDIES : SPOT SPEED STUDIES AND ITS PRESENTATION

2.6.1　Types of Speed Studies

(A) Spot Speed Studies : Spot speed measurements are mostly taken at a point (or a short section) of road way under conditions of free flow. The objective is to determine the speeds that drivers select, unaffected by the congestion. This information is used to find out general speed trends, to help in determining reasonable speed limits and to assess safety. Also they are used to determine the design speed, aid in conducting safety studies and analyze special operational situations.

(B) Travel Time Studies : Travel time is the time taken by a vehicle to cross a given segment of a street or highway to evaluate the effectiveness of traffic improvements, to provide economic analysis of alternatives and to evaluate trends in efficiency and level of service

(C) Speed Delay Studies : Delay is the time lost by a vehicle due to causes beyond the control of the driver. These studies are done to identify problem areas, to determine the efficiency of a route, to evaluate effectiveness of traffic improvements and to provide economic analyses of alternatives.

2.6.2 Spot Speed Studies and its Presentation

- When the traffic parameter over a short distance is measured, it is normally the spot speed. A spot speed is made by measuring the individual speeds of a sample of the vehicle passing a given spot on a street or highway. It is the speed of traffic at one point or spot on a traffic way (instantaneous speed). Spot speed studies are used to determine the speed distribution of a traffic stream at a specific location. The data gathered in spot speed studies are used to determine vehicle speed percentiles, which are useful in making many speed-related decisions.

(A) Application :

- Spot speed data have a number of safety applications, including the following:
    1. Speed trends: Establishing speed trends at the local, state and national level to assess effectiveness of speed limit policy.
    2. Traffic control planning: to establish speed limits, to determine safe speeds at curves
    3. Accidental analysis: For example, to determine speeds at the problem locations; to validate whether speeds are too high
    4. Geometric design, say for example, establishing the speed zone of new or existing speed limit or enforcement practices.
    5. Research studies.

(B) Factors affecting spot speed studies :

- **Driver :** Age, Gender, motive of the journey, distance of his trip;
- **Vehicle :** Type, age, weight, manufacturer and horse power;
- **Roadways and environment :** The graphical locations, grade, sight distance, number of lanes, spacing of intersections; including time of day and weather
- **Traffic :** Heavy or less volume, density, passing movements, speed regulations

(C) Methods in conducting Spot Speed Study :

Manual method:

- To observe the time required by a vehicle to cover a short distance.
- Two reference points are located at a roadway at a fixed distance apart.
- Observer starts and stops a stopwatch as vehicle enters and left the test section.
- It is very easy method.
- Disadvantage is of parallax effect.

Automatic method (radar meter detector):

- Using reflected waves of very high frequency is directed from the radar speed meter to the moving vehicle.
- The **limitations** of radar meter are:
    - The accuracy varies; they are generally +/− 1-2 mi/h.
    - When the driver slows down, this affects the results.
    - A good measurement angle must be set.
    - Multilane traffics are difficult to study.

o In heavy traffics, it is impossible to record speed of each vehicle.

2.6.3 Data Presentation and Analysis

- Data are presented in the following two ways:

2.6.3.1 Graphical Presentation

(A) Frequency Histogram :

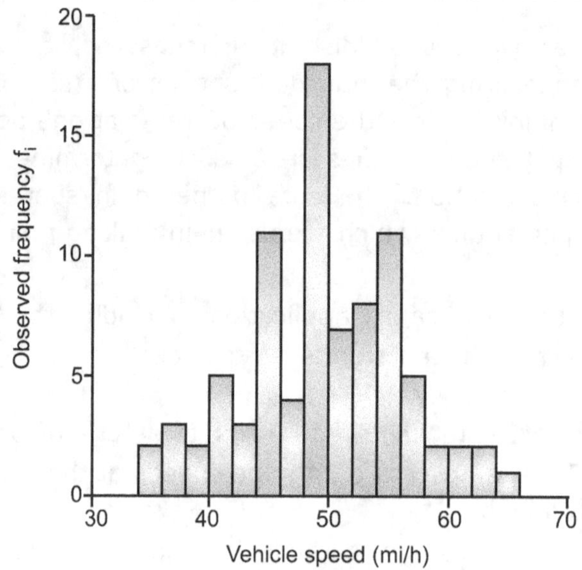

Fig. 2.8

Frequency histogram of observed vehicles' speeds

Source: Figure 2.8, Garber and Hoel (2002).

(B) Frequency distribution curve of observed vehicles' speeds :

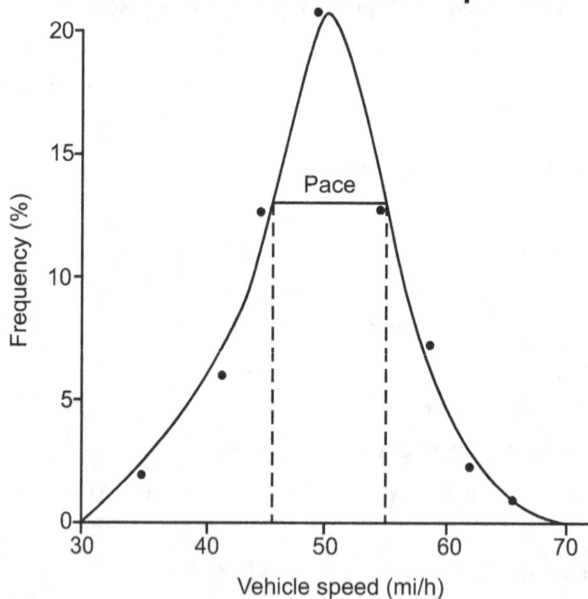

Fig. 2.9

Frequency distribution curve of observed vehicles' speeds

Source: Figure 2.9, Garber and Hoel (2002).

(C) Frequency Cumulative Curve of observed vehicles' speeds

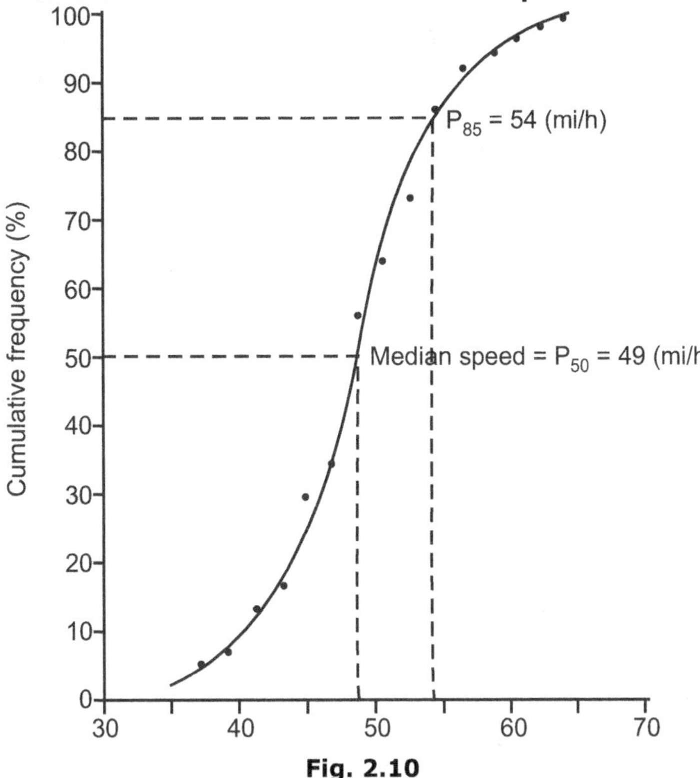

Fig. 2.10

Frequency Cumulative Curve of observed vehicles' speeds

Source: Figure 2.10, Garber and Hoel (2002).

2.6.3.2 Statistical Presentation

(A) Arithmetic mean speed :

- It is the average speed of all observed vehicles.

$$X \text{ (avg.)} = \frac{\Sigma fv}{n}$$

where ;

X (avg.) = Arithmetic mean speed

     f = frequency of observation in the particular group

     v = mean speed of each group

     n = number of observations

(B) Median speed :

- Median speed is defined as speed at which the speed distribution is separated into two equal parts. The number of the speed values observed higher than median value is equal to the number of observed speed values lower than median value of speed.

- It is a middle volume speed in the distribution of all volumes which are arranged in ascending order. It is also called 50th percentile speed (P_{50}). It means that this is the speed which occurs for 50% of the time periods in a traffic situation.

- It is very easy to understand and calculate. It is at the middle of series, and it is not affected by extreme values. Using the group frequency distribution curves, the median speed values are located graphically (determined approximately by using graph). The disadvantage is that it is very difficult to arrange large number of data values in ascending order.

(C) Modal speed :

- It is the speed value that occurs most frequently in a sample of spot speeds. Spot speeds are taken to estimate the speed distributions of vehicles in traffic at the specific location on a highway. Modal speed is a speed value that will occur frequently in the sample of spot speeds. By recording the speed of the vehicles, identification of speed characteristics for the traffic is done and the safety of the highway for the pedestrians and slow moving vehicles is assessed. Modal speed can also be computed graphically using the frequency distribution curve.

- Accuracy depends on the total amount of vehicles in the sample. Speed characteristics of the traffic are identified by the significant values such as average speed, median speed, modal speed, standard deviation of speeds, pace and i^{th} percentile spot speed.

(D) Standard deviation of speed :

- It is a measure of the spread of the individual speeds with its maximum along with minimum limits. It provides a statistical measure of speed variation at a site or road at which spot speed measurements are taken. These are calculated from standard deviation, which is as given below.

$$\sigma_s = \sqrt{\frac{\Sigma f_i(v_i - v_v)^2}{n - 1}}$$

- Value of variance is found by taking square of value of standard deviation. VAR = SD^2

- Also, it is used for comparing difference in mean speeds among spot speeds measured from two road sites. Also, it is useful for identifying any outlier speed values, which will give a hint of maximum and average speed maintained on the roads.

(E) Percentile Speeds :

- The 85th and 15th percentile speeds give an idea about the high and low speeds observed by most the drivers. The upper and lower 15% of the distribution represent speeds that are either too fast or too slow for the given conditions. These values are found graphically from the Frequency Cumulative Curve. The curve is entered on the vertical axis at values of 85% and 15%. The respective speeds are found on the horizontal axis. The 85th and 15th percentile speeds can be used to roughly estimate the standard deviation of the distribution σ_{est}, although this is not recommended when the data is available for a precise determination.

$$\sigma_{est} = \frac{v_{85} - v_{15}}{2}$$

- The 85th and 15th percentile speeds exhibit both the central tendency and dispersion of the distribution. As these values become closer to the mean, it indicates less dispersion and the central tendency of the distribution turns out to be sturdier.

- The 98th percentile speed can also be found out from the Frequency Cumulative Curve and this speed is normally used for geometric design of the road.

2.7 NEED AND METHOD OF PARKING STUDIES

- **Parking** is the act of stopping and disengaging a vehicle and leaving it unoccupied. Parking on one or both sides of a road is often permitted, though sometimes with restrictions. Some buildings have parking facilities for use of the buildings' users. Countries and local governments have rules for design and use of parking spaces. Parking is one of the major problems which is due to excessive traffic and lot of car in CBD. It is an impact of transport development and the development of a country overall. The availability of space in urban areas has increased tremendously and therefore the demand for parking space especially in areas like Central business district. Parking woes affect the mode choice too and this has great economic effects.

2.7.1 Need

1. Parking Studies are done to find out the present capacity
2. To find out the demand of the parking in the area under study and thereby to suggest any kind of improvement to it.
3. Any vehicle will at one time be parked short time or longer time, provision of parking facilities is essential
4. Need for parking spaces is usually very great in areas of business, residential, or commercial activities.
5. Nowadays, due to increased population in the urban cities, there is continuous need to facilitate public transport and adequate parking facilities.
6. Management of parking facilities in public areas like schools, colleges, hospitals and stations in the bigger cities is very essential.

2.7.2 Types of Parking Facilities/Systems

2.7.2.1 On Street Parking

- It is also known as curb facilities. The vehicles are parked on the sides of the street. Parking bays are provided alongside the curb on one or both sides of the street. Sometimes it is unrestricted, unlimited and free. Other times it is restricted parking facilities, limited to specific times for a maximum duration. And may or may not be free. Also there is handicapped parking and bus stops and loading bays. This is be generally controlled by government agencies. As per IRC, the standard dimensions of a car is taken as 5 × 2.5 meters and that for a truck is 3.75 × 7.5 meters for parking design.

Types of kerb/on-street parking :

(A) Parallel Parking : The vehicles are parked along the length of the road. As there is no backward movement involved while parking or unparking the vehicle, it is therefore the safest parking from the accident point of view. However, it uses the maximum curb length and therefore only a minimum number of vehicles can be parked for a given kerb length. This method of parking produces least obstruction to the on-going traffic on the road since least road width is used.

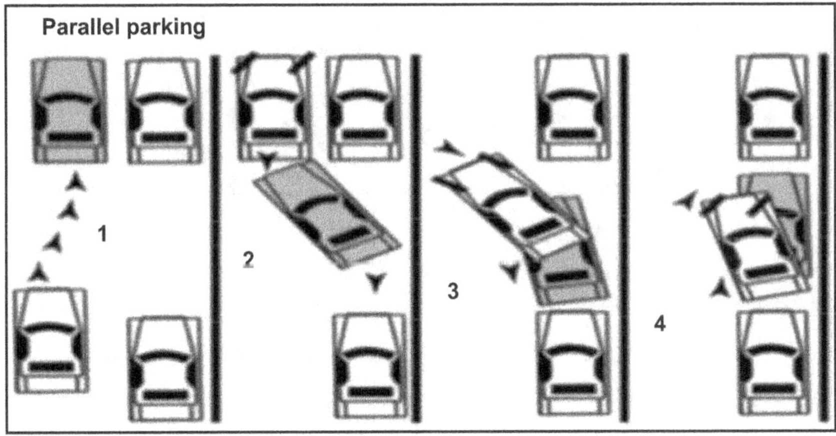

Fig. 2.11

(B) 30° parking : In thirty degree parking, the vehicles are parked at 30° with respect to the road alignment. In this case, more vehicles can be parked compared to parallel parking.

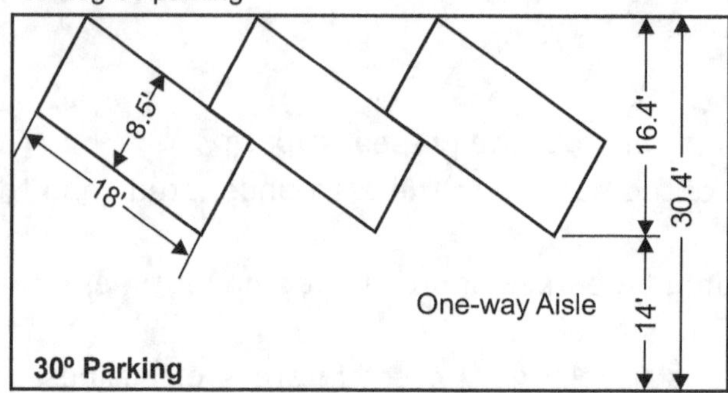

Fig. 2.12

(C) 45° parking : As the angle of parking increases, more number of vehicles can be parked. Hence compared to parallel parking and thirty degree parking, more number of vehicles can be fit into this type of parking.

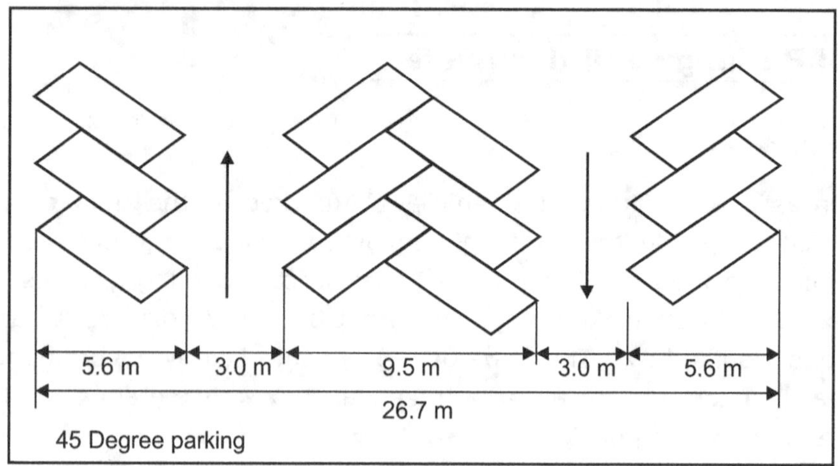

Fig. 2.13

(D) 60° parking : The vehicles are parked at 60° to the road alignment. More number of vehicles than the previous ones can be accommodated in this parking type.

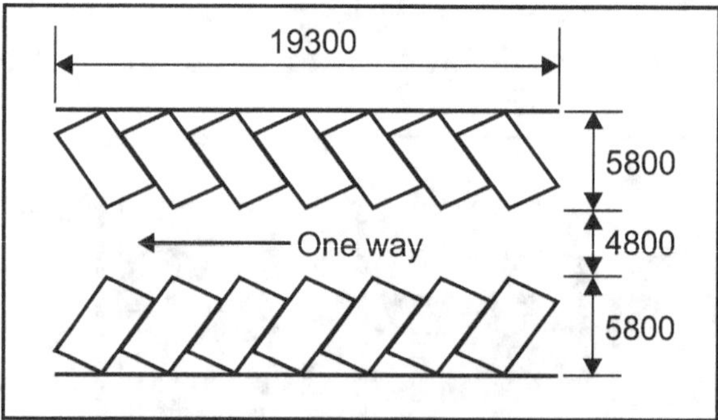

Fig. 2.14

(E) Right angle parking : It accommodates maximum number of vehicles for a given kerb length. In right angle parking or 90° parking, the vehicles are parked perpendicular to the direction of the road. Although it consumes maximum width, kerb length required is very little. In this type of parking, the vehicles need complex maneuvering and this may cause severe accidents. This arrangement causes obstruction to the road traffic particularly if the road width is less.

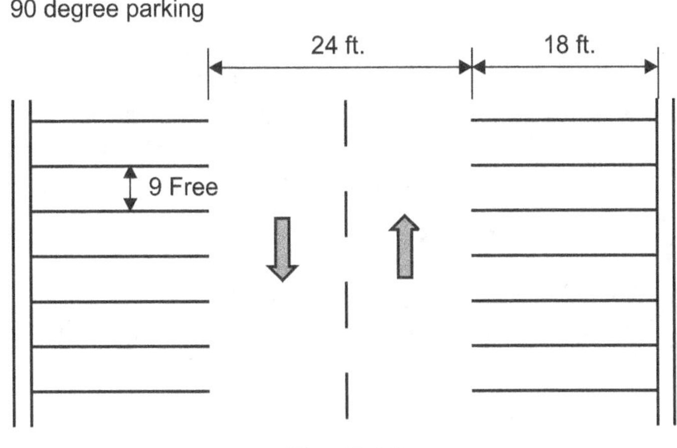

Fig. 2.15

2.7.2.2 Off Street Parking

- At many places/cities, some areas are exclusively allotted for parking which will be at some distance away from the main stream of traffic. Such a parking is known as off-street parking. They may be operated by either public agencies or private firms and are either self-parking garages or attendant-parking garages.

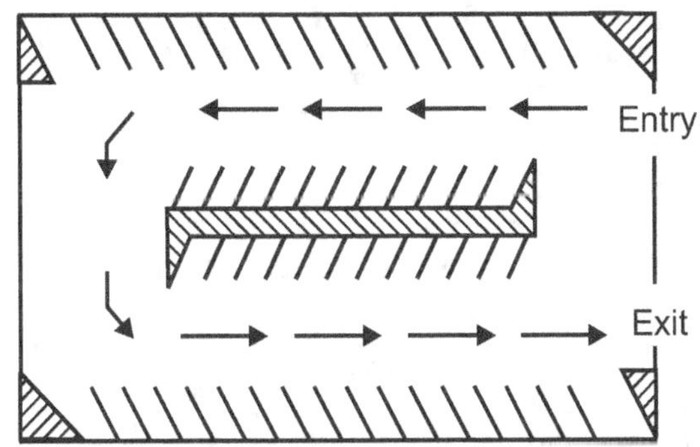

Fig. 2.16

Fig. 2.17

Fig. 2.18 : Surface car parking

Fig. 2.19 : Multistoried car parks

Fig. 2.20 : Roof parks

Fig. 2.21 : Mechaincal parks

Fig. 2.22 : Underground car parks

2.7.3 Parking Surveys/Studies

- Parking surveys are conducted to calculate the parking statistics. The most common parking surveys conducted are in-out survey, fixed period sampling and license plate method of survey.

2.7.2.2 Off Street Parking

- Before taking any steps for the improvement of parking conditions, data regarding availability of parking space, extent of its usage and parking demand are required. Parking fares are also to be estimated and fixed. Parking surveys are done to provide all this information. As the duration of parking varies with different vehicles, many statistics are in practice to access the parking needs. The following parking statistics are generally found out:

(A) Parking accumulation : It is defined as the number of vehicles parked at a given instant of time. Normally this is expressed by accumulation curve. Accumulation curve is the graph obtained by plotting the number of bays occupied with respect to time.

(B) Parking volume : Parking volume is the total number of vehicles parked in a given duration of time. This does not account for repetition of vehicles. The actual volume of vehicles entered in the area is recorded.

(C) Parking load : Parking load gives the area under the accumulation curve. It can also be obtained by simply multiplying the number of vehicles occupying the parking area at each time interval with the time interval. It is expressed as vehicle hours.

(D) Average parking duration : It is the ratio of total vehicle hours to the number of vehicles parked.

(E) Parking turnover : It is the ratio of number of vehicles parked in a duration to the number of parking bays available. This can be expressed as number of vehicles per bay per time duration.

(F) Parking index : Parking index is also called occupancy or efficiency. It is defined as the ratio of number of bays occupied in a time duration to the total space available.

2.7.3.2 Parking Surveys

- Parking surveys are done to collect the data to assess the above mentioned statistics and these are as follows:

    (A) In-out survey : In this survey, the occupancy count in a pre-decided parking lot is taken at the beginning. Then the number of vehicles that enter the parking lot for a fixed time interval is counted. The number of vehicles that leave the parking lot is also counted. The final occupancy in the parking lot is also ascertained. It is a less labor intensive, cheaper and quick method. Only one person may be enough. Not all the statistics can be calculated by this survey/study.

    (B) License plate method of survey : This survey gives the most accurate and realistic data. In this survey, every parking stall is monitored at a continuous interval of 15 minutes or so and the license plate number is noted down. This gives the data regarding the duration for which a particular vehicle used the parking bay. This facilitates in calculating the fare because fare estimation is based on the duration for which the vehicle was parked. This method is very labour intensive and costly.

Important Points

- Traffic study is a planning tool/method used by transportation/Traffic engineers to predict demands on the transportation network and subsequently to determine transportation improvements.

- Types of Traffic Studies :
  - Traffic volume study
  - Speed and delay study
  - Origin and destination study
  - Traffic flow characteristics study
  - Traffic capacity and level of service study
  - Parking study
  - Accident study
- Traffic volume studies are conducted to determine the number, movements, and classifications of roadway vehicles at a given location.
- A **traffic count** is a count of vehicular or pedestrian traffic, which is conducted along a particular road, path, or intersection.
- Traffic volume study is the quantity of vehicles crossing a section of road per unit time at any selected period. It is usually expressed in terms of Passenger Car Unit (PCU).
- There are three important cyclical variations:
  - (1) Hourly pattern   (2) Daily Pattern
  - (3) Monthly and yearly Pattern:
- The flow data collected by a traffic/transportation engineer observed on field can be classified into three types depending upon the length of observation:
  - Measurement at a point
  - Measurement over a short section
  - Measurement over a long section
- The most important factor for a successful survey is the level of participation by public upon which the success of any method depends.
- Types of Speed Studies :
  - Spot Speed Studies
  - Travel Time Studies
  - Speed Delay Studies
- **Application :**
  - Speed trends
  - Traffic control planning
  - Accidental analysis
  - Geometric design
  - Research studies

Practice Questions

1. Define the following terms : (a) Traffic Study, (2) Public Transport Survey.

2. Enlist the factors affecting spot speed study.

3. Write note on Road side interview.

4. Define : Peak hour factor.

5. State types of volume measurement.

6. Write note on Parking Surveys.

7. Enlist Safety applications of Spot Speed Studies.

8. State and explain types of Speed Studies.

9. Explain Passenger Car unit.

10. Explain detail methods of collection of Traffic Volume Count Data.

ROAD SIGNS AND TRAFFIC MARKINGS

Objectives

3a.   Interpret the traffic signs at the given road intersection or road.

3b.   Suggest the road signs for given traffic situation.

3c.   Explain the necessity of pavement markings on given road.

3d.   Plot the pavement markings as per norms.

3.1 TRAFFIC CONTROL DEVICES

3.1.1 Definition

- The various aids and devices used to control, regulate and guide traffic may be called traffic control devices.

- The main object of providing traffic control devices on road is to provide safe, convenient and economical transportation of persons and goods.

- The general requirements of traffic control devices are : attention, meaning, time for response and respect of road users.

3.1.2 Necessities/objectives/uses/functions of Traffic Control Devices

1.   They give information of routes, directions.

2.   They give information about points of intersections.

3.   They give timely warning of hazardous situations.

4.   They help to regulali the traffic.

5.   They are useful for traffic control management.

6.   They are useful for safe and smooth traffic flow.

3.1.3 Types of Traffic Control Devices

- Traffic control devices:
  - (a) Road signs,
  - (b) Road markings,
  - (c) Road signals and
  - (d) Traffic Island
- In addition to these, road lights are also useful in guiding traffic during night.

3.2 ROAD SIGNS

3.2.1 Definition of Road Signs

- To provide safety to the traffic, to expedite the traffic, to control the traffic and to guide the traffic road signs are provided.
- Road signs are the devices in the form of symbols and inscriptions mounted on fixed or portable support provided on the roads to give information, warning or guidance to the vehicles and road users.

3.2.2 Objects/Necessity/Purpose/Funtions of Road Signs

- The purposes of road signs can be summarized as:
  - (a) To regulate the traffic by imparting messages to the driver when to stop, give way or limit their speeds.
  - (b) To give timely warning of hazardous situations when they are not self-evident.
  - (c) To supply information on highway routes, directions and points of interest.
  - (d) To provide safety to traffic.
  - (e) To control the traffic.
  - (f) To guide the traffic.

3.3 CLASSIFICATION OF ROAD SIGNS (AS PER IRC : 67 – 1977)

- According to Indian vehicle acts and as per **IRC: 67-1977 code of practice**, road signs are classified as:
  1. Regulatory Signs (Mandatory Signs).
  2. Warning Signs (Cautionary Signs). .... (W-15)
  3. Informatory Signs (Guiding Signs).

3.3.1 Regulatory Signs (Mandatory Signs)

1. **Regulatory Signs (Mandatory Signs):**
   - o These signs are mandatory signs and inform the road users about certain laws and regulations to provide free flow of traffic. These road signs include all such signs, which give notice of special obligations, prohibitions or restrictions with which the road users must comply. Violation or non-compliances of these regulations is a legal offense.
   - o The size, shape and details of all traffic signs have been standardized by the Indian Road Congress (IRC).
   - o These regulatory signs are provided with a 600 mm diameter red disc and are installed at a height of 2800 mm above the ground level upto the centre of the plate. This post also carries a rectangular definition plate below the circular disc.

○ The regulatory signs are classified under the following sub-heads:

(i) Stop and Give-way Sign:

- This sign is used on roads where traffic is required to stop before entering the major road. These signs are usually used on minor roads at their intersection with a major road where conditions are usually hazardous due to restricted visibility or bad alignment etc., making it imperative to the minor road traffic to stop.

- This sign is octagonal in shape and red in colour with white border. This sign may be used in combination with a rectangular definition plate with the word 'STOP' written in English and other languages if necessary.

- The Give-way sign is used to control the vehicles on a road so as to assign right of way to traffic on other roadways. This sign is in equilateral triangle shape with the apex downwards and white in colour with a red border.

- The length of normal size and small size is 90 cm and 60 cm including border respectively. The size of border is 7 cm and 4.5 cm for normal and small size respectively. Fig. 3.1 shows Stop and Give-way signs.

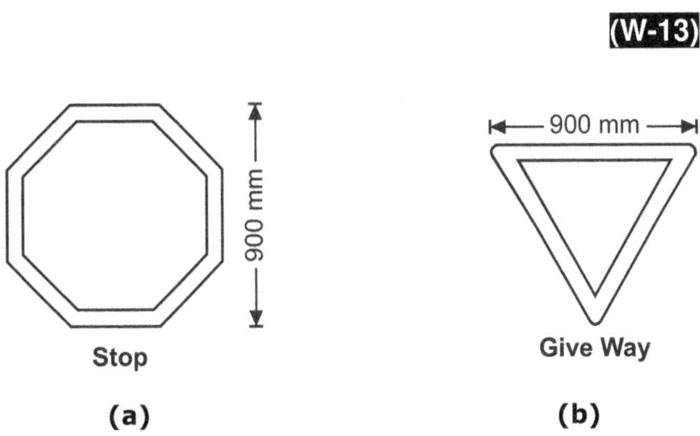

Fig. 3.1: Stop and Give-Way Signs

(ii) Prohibitory Signs:

- These signs are meant to prohibit certain traffic movements. These are circular in shape with white background and red borders, black symbols and red oblique bar. The Fig. 3.2 shows a typical structure of Prohibitory Sign post.

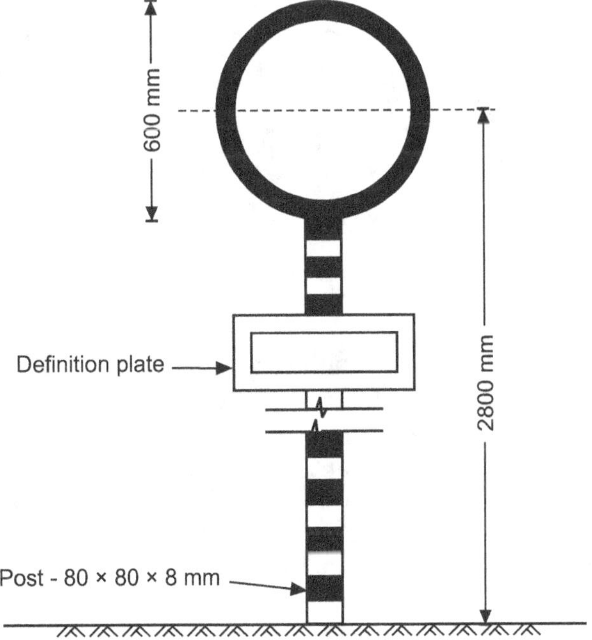

Fig. 3.2 : A Typical Structure of Prohibitory Sign Post

- The diameter of normal size and small signs are 60 cm and 40 cm with border width 6.5 cm and 4.5 cm respectively. The width of oblique bar is kept 6.0 cm and 4.0 cm for normal and small sized signs.

Fig. 3.3 : Prohibitory Signs

- The common prohibitory signs are Straight prohibited, No entry, One way, Vehicles prohibited in both directions, All Motor Vehicles are prohibited, Truck prohibited, Bullock Cart and Hand Cart prohibited, Cycle prohibited, Pedestrians prohibited, Right/Left Turn prohibited, U-turn prohibited, Horn prohibited, Overtaking prohibited etc. The Prohibitory Signs are as shown in Fig 3.3.

(iii) No Parking and No Stopping Sign:

- This sign meant to prohibit parking of vehicles at that place; the definition plate may indicate the parking restriction with respect to days, distance etc. The parking sign is circular in shape with a blue background, a red border and an oblique red bar at an angle of 45°. No Stopping/Standing sign meant to prohibit stopping of vehicles at that place; the scope of the prohibition may be indicated on the definition plate.

- The No Stopping/Standing sign is circular in shape with blue background and two oblique bars at 45° and right angles to each other. Fig. 3.4 shows No Parking and No Stopping/Standing sign.

(S-14, 15)

Fig. 3.4 : No Parking and No Stopping/Standing Sign   **(S-09; W-11)**

(iv) Speed Limit and Vehicle Control Sign:

- These signs are meant to restrict the speed of all or certain classes of vehicles on a particular stretch of road. These signs are circular in shape and have white background, red border and black numerals indicating the speed limit.

(W-13, S-14, 15)

Fig. 3.5 : Speed Limit and Vehicle Control Sign   **(S-09; W-11)**

- The vehicle control signs are also similar to speed limit signs with black symbols instead of numerals. The common control signs are Width limit, Height limit, Length limit, and Load limit. The definition plate may be used in combination to give more details, symbolically or by wards. Fig. 3.5 shows Speed Limit and Vehicle Control signs.

(v) Compulsory Direction Control Signs:

- These signs are indicated by arrows, the appropriate directions in which the vehicles are obliged to proceed, or the only directions in which they are permitted to proceed. These signs are circular in shape with blue background and white direction arrows. The common Compulsory Direction Control signs are Compulsory Turn Left, Ahead Only, Ahead or Turn Left/Right and Keep Left. Fig 3.6 shows Compulsory Direction Control Sings.

(S-15)

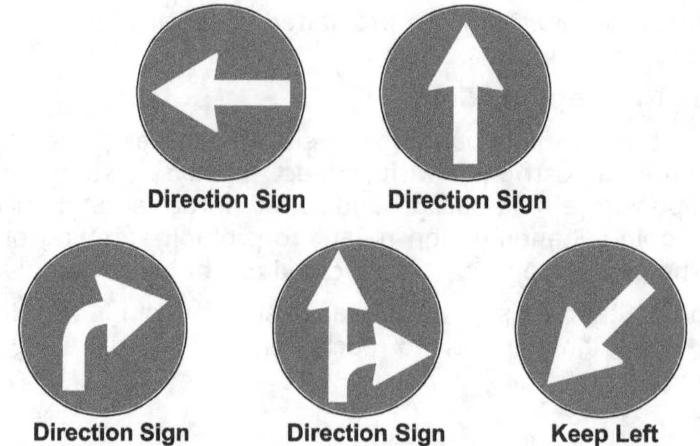

Fig. 3.6 : Compulsory Direction Control Signs

(vi) Restriction Ends Signs:

- These signs indicate the point at which all prohibitions notified by prohibitionary signs for moving vehicles ceases to apply. These signs are also circular with a white background and a broad diagonal black band at 45°. Fig. 3.7 shows restriction end sign.

Fig. 3.7 : Restriction End Sign

3.3.2 Warning Signs (Cautionary Signs)

- These signs are used to warn the road users of the hazardous conditions either on the road or adjacent to it. The warning signs are in the shape of equilateral triangle with its apex upwards. They have a white background, red border and black symbols.

- These signs are located at sufficient distance in advance of the hazard warned against; these distances are 120, 90, 60 and 40 metres respectively on National/State Highways, Major District Roads, Other District Roads and Village Roads. On the urban roads, this distance is 50 metres.

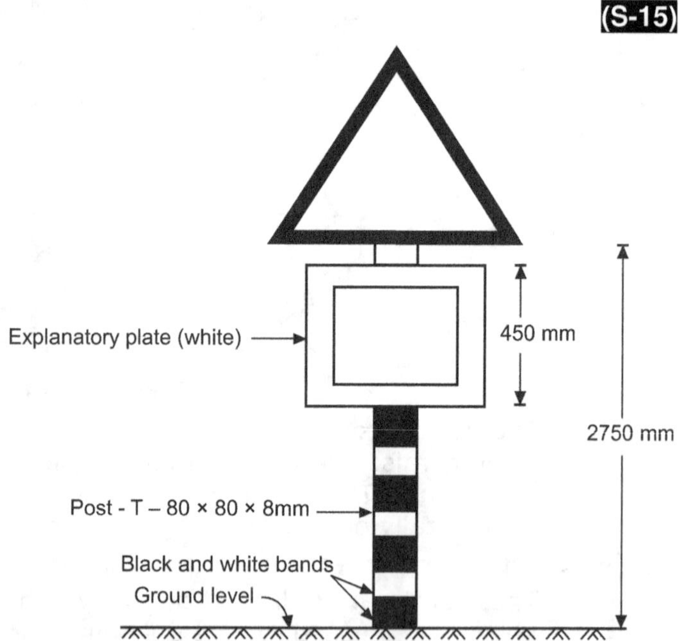

Fig. 3.8 : A Typical Structure of Warning Signs

- The commonly used warning signs are Right Hand Curve, Left Hand Curve, Right/Left Hair Pin curve, Right/Left Reserved Bend, Steep Ascent/Descent, Narrow Bridge Ahead, Narrow Road

Ahead, Gap in Median, Slippery Road, Cycle Crossing, Pedestrians Crossing, Men at Work, Cross Road, Side Road, T-Intersection, Y-Intersection, Rough Road, Unguarded railway Crossing, Guarded railway Crossing, School, Cattle, Falling Rocks, Staggered Intersection etc. Fig. 3.9 shows Warning Signs.

Right Hand Curve

Left Hand Hairpin Bend

Right Hand Hairpin Bend

Right Hand Curve

Falling Rocks

Men at Work

Roundabout

(Unguarded)

(Guarded)

Railway crossing

Zig-zag Road

Dangerous Ascent

Dangerous Descent

Narrow Road Ahead

Cattle

Slippery Road

Loose Gravel

Pedestrian Crossing

School

Refreshment place   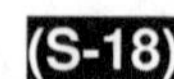

Irregular Intersections

Intersection

Intersection

Irregular Intersections

Staggered Intersection

Staggered Intersection

Cross Road

Major Road Ahead

Cross Road

Side Road

Side Road

Uneven Road

Cycle Crossing

Major Road Ahead

Narrow Bridge

Gap in Median

Dangerous Dip

Speed Breaker

Fig. 3.9 : Warning Signs

Informatory Signs (Guiding Signs):

- These signs are used to guide the road users along routes, inform them of destination, distance and provide with information to make travel easier, safe and pleasant. These signs are grouped under the following sub-heads:

(i) Direction and Place Identification Signs:

- These sings are rectangular with white background, black border, black arrows and letters. The inscription should be in English and other languages as necessary. The common signs of this group include Destination sign, Direction sign, Re-assurance signs, Route Marker, and Place Identification signs. Fig. 3.10 shows some Informatory signs.

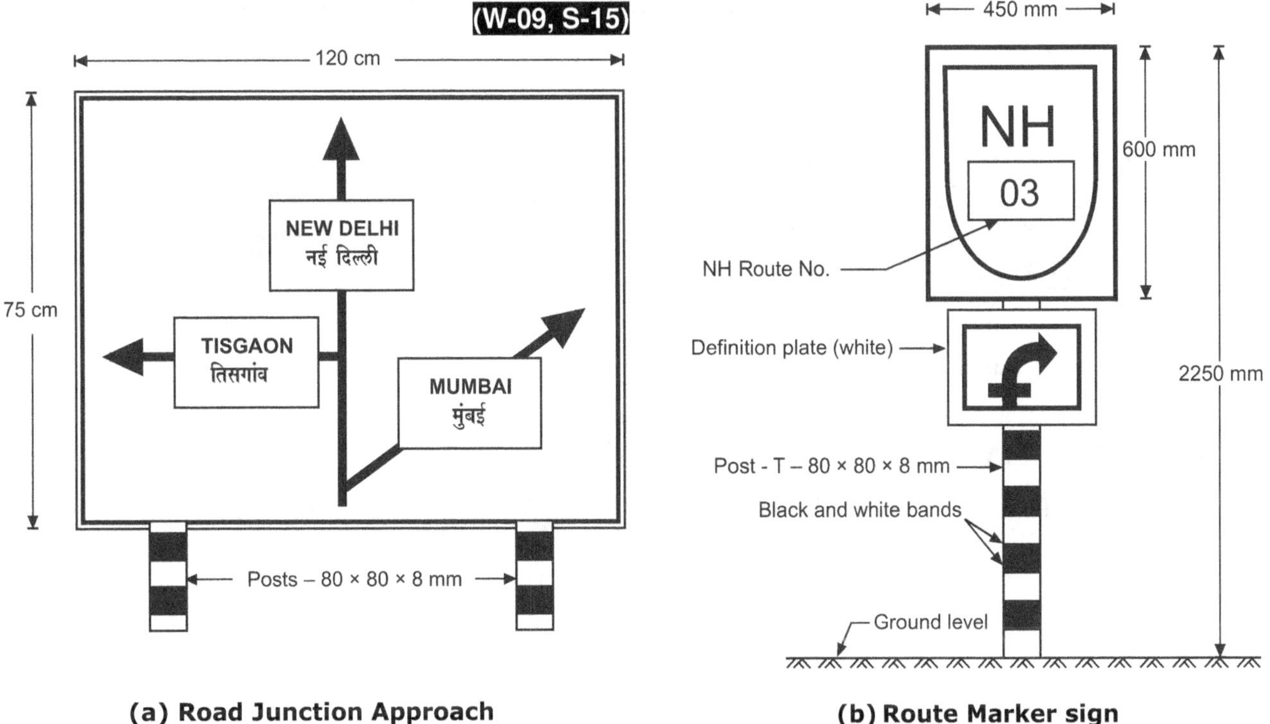

(a) Road Junction Approach   **(b) Route Marker sign**

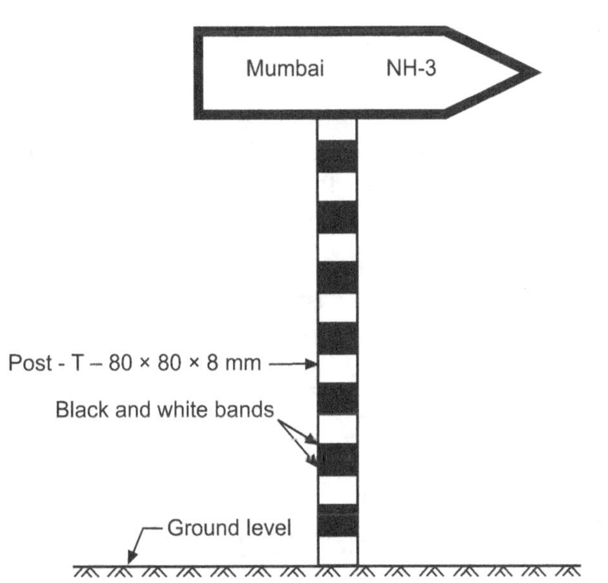

(c) Direction Marker

Fig. 3.10: Informatory Signs

(ii) Facility Information Signs:

- These signs are rectangular with blue background and white/black letters/symbols. The most common Facility Information signs are Petrol Pump, Public Telephone, Hospital, First Aid Post, Food Place and Resting Place. Fig. 3.11 shows Facility Information signs.

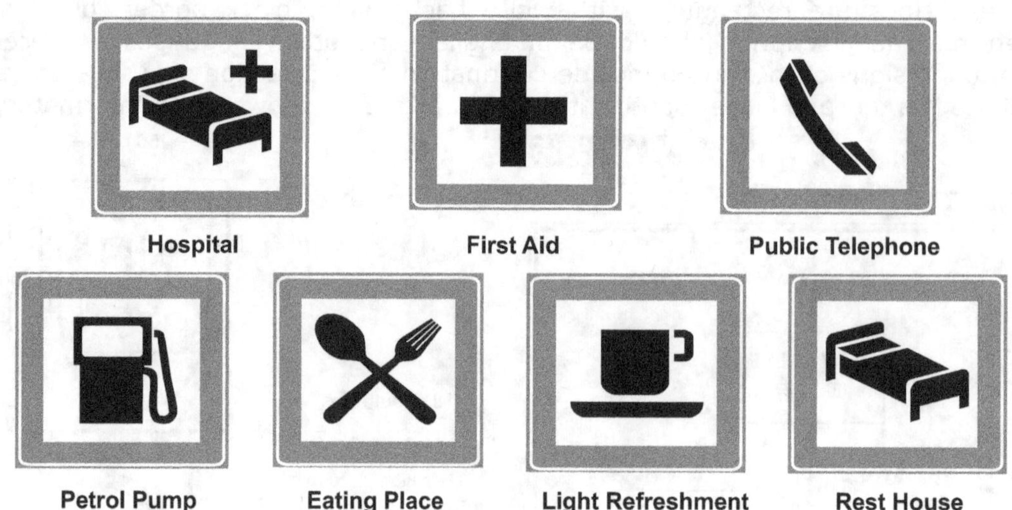

Fig. 3.11: Facility Information Signs

(iii) Other Useful Information Signs:

- These signs include No Through Road, No Through side Road etc. Fig 3.12 shows other useful signs.

Fig. 3.12: No Through Road, No Through Side Road

(iv) Parking Signs:

- Parking signs are set-up parallel to the road using square sign boards with blue background and white coloured letter 'P'. In same signboard additional plate may be used to indicate category of vehicle for which parking space is reserved, direction of parking space etc. Fig. 3.13 shows Parking signs.

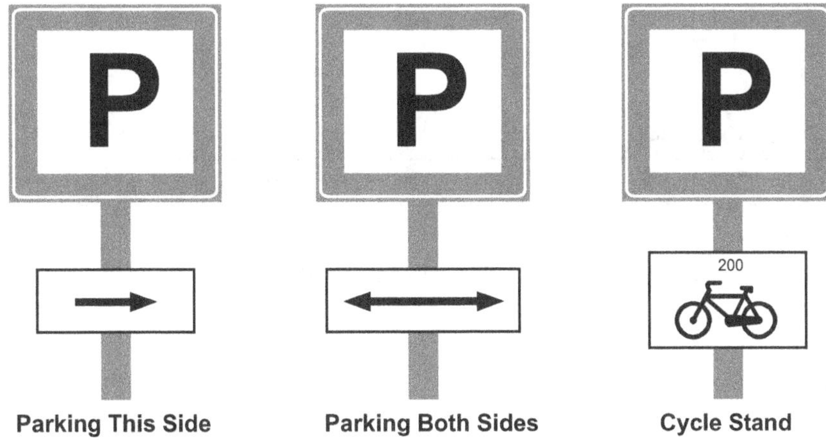

Fig. 3.13: Parking signs

(v) Flood Gauge Signs:

- These signs should be stalled as all cause ways and submersible bridges or culverts to indicate to the road users the height of the flood above road level.

3.3.3 Location of Warning Signs

- In India, the sign should be located on the left side of the road.

- On multi-way carriageways they may be repeated on both sides.

- On wide expressways overhead signs are also provided.

- Warning signs are required to be located at the following distances
  (a) Urban location : 50 m
  (b) Non-urban location :

Types of Road	Plain and Rolling Terrain	Hilly Terrain
(1) NH/SH	120 m	60 m
(2) MDR	90 m	50 m
(3) ODR	60 m	40 m
(4) V.R	40 m	30 m

3.3.4 Points to be Considered While Design the Road Sign

(1) Material : As per IRC : 67-2010/BIS/ASTM

(2) Size of Signs :
   (i)   Small size (600 mm)
   (ii) Normal size (900 mm)
   (iii) Large size (1200 mm)

(a) The size refers to the diameter of the circular signs

(b) Height of octagonal signs

(c) Side of triangular signs

(3) Foot :
   (i)   For small size (100 mm)
   (ii) For normal size (150 mm)
   (iii) For large size (225 mm)

(4) Location :
   (i)   Small size : (Minor Roads)
   (ii) Normal size : (Highway, Urban Road)
   (iii) Large size : (Express Way)

(5) Design speed :
  (i)   Small size : 60 kmph
  (ii) Normal size : 60 to 100 kmph
  (iii) Large size : more than 100 kmph

3.3.5 Points to be Considered While Erecting the Road Signs

(1) Material : As per BIS/ASTM/IRC
(2) They should be placed such that they could be seen easily.
(3) They could be recognized easily.
(4) They could be seen and recognized in time.
(5) For road with Kerbs :
     Transver location of sign → edge of the sign adjacent to the road is not less than 0.6 m away from the edge of the Kerbs.
(6) For road without Kerbs (Rural road) the nearest edge may be 2.0 to 3.0 m from the edge of the carriageway.
(7) The signs should be mounted on sine posts.
(8) Sign posts should be painted alternately with 250 mm white and black bands.
(9) The size, shape, colour code and symbols should be as per IRC.
(10) Location of signs of should be as per IRC.
(11) The reverse side of all sign plates should be painted in grey colour.

3.4 TRAFFIC MARKING

3.4.1 Definition

Traffic Markings :

- Traffic markings are the special signs used to regulate the traffic. Traffic markings are made of lines, patterns, words, symbols or reflectors on the pavement, Kerb, and side of islands or on fixed objects within or near the roadway.

- The markings are made using paints in contrast colour of the background. Generally, light reflecting paints are used for traffic markings.

- The longitudinal lines should be at least 10 cm thick and the transverse lines should be of such thickness that they are visible from a sufficient distance to give road users adequate time to take proper action.

3.4.2 Objects of Traffic Markings

(1) To warn the road users
(2) To inform the road users
(3) To guide the road users

3.5 CLASSIFICATION OF TRAFFIC MARKINGS

3.5.1 Traffic Markings

- The traffic markings may be classified into the following categories.

  (a) Pavement markings (Carriage way markings)

  (b) Kerb markings

  (c) Object markings

  (d) Reflector unit markings.

3.5.2 Pavement Markings

- Pavement or carriageway markings may generally be of white paint. Yellow colour markings are used to indicate parking restrictions and for the continuous centerline and barrier line markings. Pavement markings are made for the following purposes;

  (i) **Centreline:** To divide the roadway, centerline is marked. These lines are used to separate two-way traffic. On rural highway with two or three lanes, single broken lines of width 0.1 m and length 4.5 m segments and 7.5 m gaps may be painted on straight stretches of NH and SH; these may be decreased to 3.0 and 6.0 m at horizontal curves and approaches to intersection. On other roads at straights the segments are 3.0 m in length and gaps 6.0 m (which are reduced to 3.0 m at curves and approaches to intersection). On four or six lane undivided two solid continuous parallel lines of 0.1 m width with 0.05 to 0.10 m space in between are painted.

  (ii) **Lane line:** These lines indicate traffic lanes. These are used to guide the traffic. These lines are drawn in same way as that of centerlines are drawn.

  (iii) **No passing zone (No-overtaking zone) markings:** These markings are made to indicate the overtaking is not permissible.

  (iv) **Stop lines:** These are made near the pedestrian's crossings, signalized intersections etc. to indicate that vehicle have to stop before this line and proceed.

  (v) **Cross walk line:** These markings are made, where the pedestrians have to cross the pavement.

  (vi) **Parking space limit:** These markings are made to indicate the space for parking and its proper utilization.

  (vii) **Pavement edge lines:** These lines are used to indicate the edges of carriageways, which have no Kerbs. Pavement edge lines serves as a visual guidance for the drivers, indicating to them the limits upto which the driver can safely venture. They are especially useful during adverse weather and poor visibility.

  (viii) **Carriageway width reduction transition marking:** These markings are used when there is reduction in lanes and hence reduction in carriageway width.

  (ix) **Obstruction approach markings:** These markings are necessary to guide traffic on the approach to fixed obstruction within the carriageway. The marking must be designed to guide the traffic away from the obstruction.

  (x) **Cyclist crossing:** The cyclist to cross the road uses these markings. Cyclist crossing should have the cycle tracks. They are marked by means of squares 50 cm × 50 cm at a center-to-center spacing of 100 cm.

  (xi) **Route direction arrows:** Route direction arrows are used to guide effectively the traffic in the correct direction and lanes. Because of low angle at which such markings are viewed by the drivers, these markings must be elongated in the direction of traffic to be properly legible.

  (xii) **Word messages:** Word messages are used to convey information to guide, warn or regulate traffic. Some of the common messages used are: STOP, SLOW, SPEED - 25 etc.

(xiii) Markings at approaches to intersections: For orderly and guided traffic movement at intersections, markings can be profitably employed. Markings at intersections can be a combination of centre lines, urn markings, lane markings, stop lines, route direction arrows etc.

(xiv) Bus stops: These markings are done at the bus stops to guide the pedestrians those who are using the buses as a mean of transportation.

3.5.3 Kerb Markings

- The traffic markings provided to the Kerbs of object lying within the carriage way are Kerbs marking. These markings are provided to all islands, dividing strips of carriage way and to Kerbs directly ahead of traffic at T-junctions etc. These may indicate certain regulations like, parking regulations. Also the markings on the Kerb and edges of island with alternate black and white lines increase the visibility from a long distance.

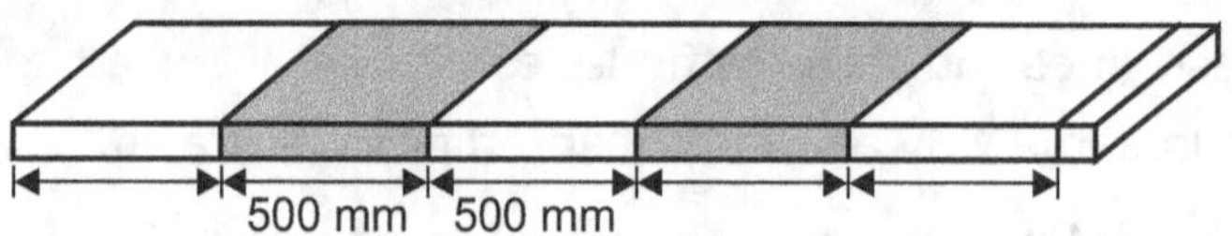

Fig. 3.14

3.5.4 Object Markings

- Support of bridges, level crossing gates, traffic island, narrow bridge etc. some times cause obstruction in visibility and cause accidents. Hence, to increase visibility, these objects should be properly marked with alternate band of white and black strips.

3.5.5 Reflector Unit Markings

- Reflector markers are used as hazard markers and guide markers for safe driving during night. Hazard markers reflecting yellow light should be visible from a long distance of about 150 m. These are provided at Kerbs to increase their visibility, especially during night hours.

3.5.6 Types of Road Markings

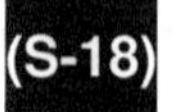

Types of road markings:

- Various types of road markings are classified as given below:
  - **Carriage way markings:** Longitudinal markings such as centre line, traffic lanes, border or edge lines, bus lane etc. and 'no parking zones', 'warning lines' etc.
  - **Markings at intersections:** Stop lines, pedestrian crossings, direction arrows, give way, marking on approaches to intersection, speed change lanes, box marking etc.
  - **Marking at hazardous locations:** Obstruction approaches, carriageway width transition, road-rail level crossings, check barriers etc.
  - **Marking for parking:** Parking space limits, parking restrictions, bus stops etc.
  - **Word messages:** Stop, slow, bus, keep clear, right turn only, exit only etc.

Object markings: Kerb marking, objects within the carriageway, objects adjacent to the carriageway etc.

3.5.7 Colour of Road Markings

1. **Alternate bands of white and black :** It is used for Kerb and object markings.

2. **Yellow colour :** It is used for making for parking restrictions. It is used for continuous centre and barrier line markings.

3. **White :** All carriage way markings.

3.5.8 Center Line Markings

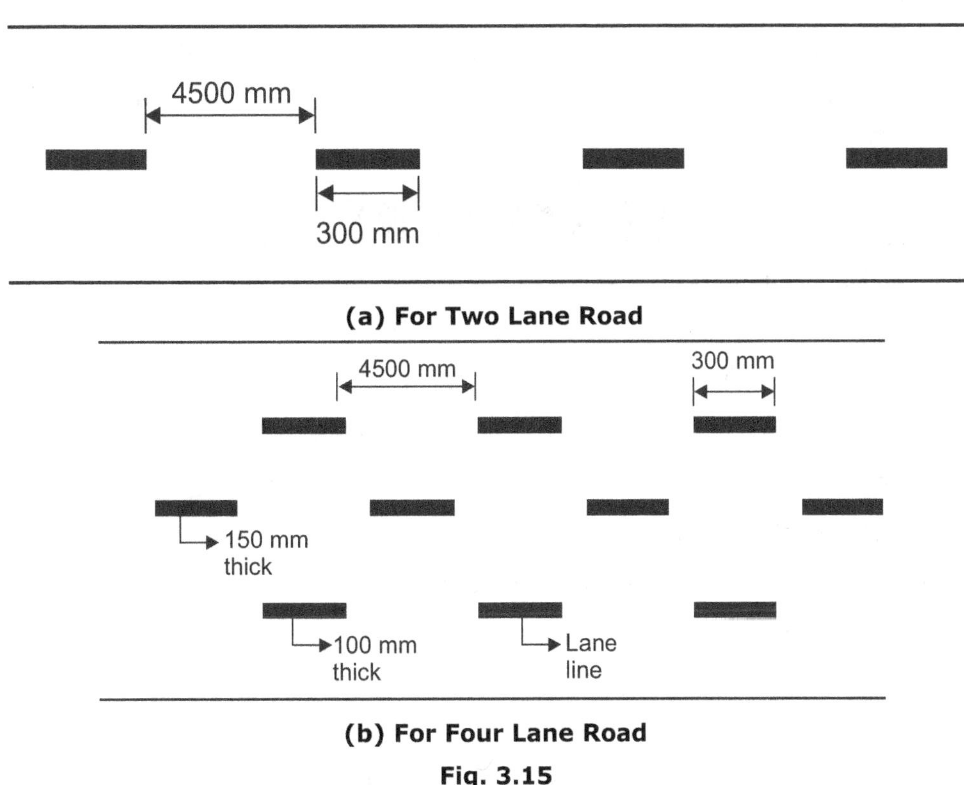

(a) For Two Lane Road

(b) For Four Lane Road

Fig. 3.15

Important Points

- The various aids and devices used to control, regulate and guide traffic may be called traffic control devices.

- Traffic control devices:

   (a) Road signs,

   (b) Road markings,

   (c) Road signals and

   (d) Traffic Island

- Road signs are the devices in the form of symbols and inscriptions mounted on fixed or portable support provided on the roads to give information, warning or guidance to the vehicles and road users.

- Road signs are classified as:

   (1) Regulatory Signs (Mandatory Signs)

   (2) Warning Signs (Cautionary Signs)

   (3) Informatory Signs (Guiding Signs)

- The size, shape and details of all traffic signs have been standardized by the Indian Road Congress (IRC).

- The common prohibitory signs are Straight prohibited.

- This sign meant to prohibit parking of vehicles at that place; the definition plate may indicate the parking restriction with respect to days, distance etc.

- **Warning Signs (Cautionary Signs) :** These signs are used to warn the road users of the hazardous conditions either on the road or adjacent to it. The warning signs are in the shape of equilateral triangle with its apex upwards. They have a white background, red border and black symbols.

- Parking signs are set-up parallel to the road using square sign boards with blue background and white coloured letter 'P'.

- Traffic markings are the special signs used to regulate the traffic. Traffic markings are made of lines, patterns, words, symbols or reflectors on the pavement, Kerb, and side of islands or on fixed objects within or near the roadway.

Practice Questions

1. Give names of different traffic control devices.

2. Give any four objects of Road signs.

3. Give classification of Road signs as per IRC.

4. Explain about location of warning signs in urban and no urban areas.

5. Define Traffic markings.

6. Give classification of traffic markings.

7. Give any four points to be considered while erecting the road signs.

8. Give any four points to be considered while designing the road signs.

9. Draw a sketch of Kerb marking and centre line markings.

Chapter 4

TRAFFIC SIGNALS AND TRAFFIC ISLANDS

Syllabus

4.1 **Traffic Signals :** Definition

4.2 **Types of signals :** Traffic control signals, pedestrian signals, special type of traffic signals

4.3 Advantages and disadvantages of traffic control signals

4.4 **Types and traffic control signals :** Fixed time, manually operated, traffic actuated signals

4.5 Location of signals

4.6 Compute signal time by fix time cycle, trail cycle, approximate, Webster's and IRC method and sketch timing diagram for each face.

4.7 **Traffic islands :** definition, advantages and disadvantages of providing islands.

4.8 **Types of traffic islands :** rotary or central, channelizing or refuge

4.9 **Road intersections or junctions :** Definition, Types

4.10 **Intersection at grade :** basic requirements of good intersection at grade, types

4.11 **Grade separated intersection :** advantages and disadvantages, types-over pass or flyovers-Cloverleaf pattern, trumpet type, underpass

Objectives

4a Justify the necessity of traffic signals at the given inter section of a road.

4b Explain the principle of co-ordinated signals, on road.

4c Categorize the traffic island existing at the given road intersection.

4d Suggest the relevant measures to guide the traffic in the given situation.

4.1 TRAFFIC SIGNALS

Definition :

- The devices used for controlling, warning or guiding the traffic are defined as traffic signals.

- Traffic signals are provided at intersection of roads to control and guide the traffic especially heavy vehicular traffic.

4.2 TYPES OF SIGNALS

- Traffic signals are of following types :

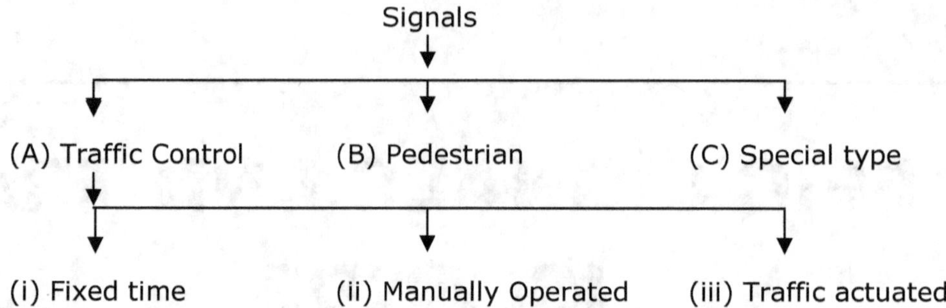

(A) Traffic Control Signals :

- The signals directing the traffic to stop and then allowing the same to go by their indications are termed as traffic control signals.
- Such signals are comprised of three coloured lights i.e., red, yellow and green.
- It faces the traffic flow and indicates the colour to guide them.
- They are generally operated by electric supply.
- The red light indicates the traffic to stop, the yellow light indicates the clearance time for moving vehicles to clear off the area of intersection and green light indicates the traffic to go.
- Some traffic control signals are also provided with additional green light to give turning directions to traffic flow.

Advantages :

- Following are the advantages of providing traffic control signals :

    (i)   Orderly movement of traffic is facilitated through traffic control signals.

    (ii) The frequency of right angled accidents gets reduced through these signals.

    (iii) They provide authority to drivers to move with confidence.

    (iv) They directs traffic with less congestion on different vehicular routes.

    (v) Safe interception of heavy traffic at road intersection is achieved through such signals.

    (vi) The speed of vehicle is maintained on main and secondary roads.

Disadvantages :

- Following are the disadvantages of providing traffic control signals :

    (i)   Sometimes these signals cause delay in quick movement of heavy traffic.

    (ii) These signals may result re-entrant collision of vehicles.

    (iii) Improper design and location of signals may lead to violations of control system.

    (iv) Electric power failure or any other defect may lead to confusion to traffic movement.

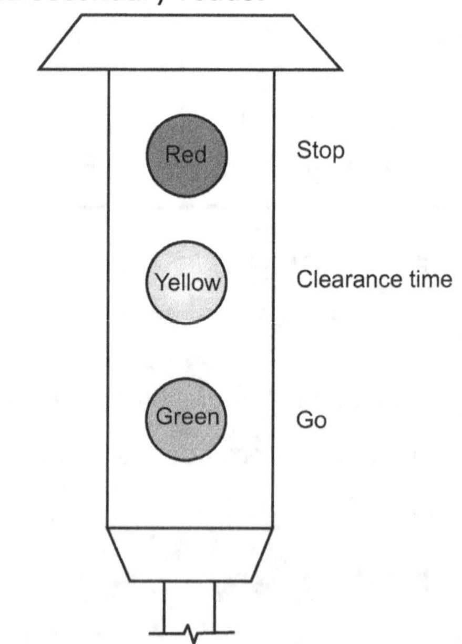

Fig. 4.1 : Traffic Control Signal Head

(B) Pedestrian Signals :

- The signals directing the pedestrians to safe crossing of roads are termed as pedestrian signals.
- Such signals are installed at intersection of roads with heavy vehicular traffic.
- These signals and their timings are guiding pedestrians to cross the roads safely.
- Such signals are interlinked with the traffic control signals for safe operation.
- Urban roads are provided with appropriate warnings and informatory signs.

(C) Special type traffic Signals :

- Special traffic signals like 'flashing beacons' are provided at certain critical areas to warn the traffic.
- Flashing beacons are having red and yellow light in each face which is illuminated by rapid intermittent flashes.
- Flashing red light means stop signal. A driver should stop before entering the nearest cross-walk at an intersection or at a stop line when marked.
- Flashing yellow light means caution signal. The drivers of vehicles may proceed through the intersection or pass such signal only with caution.

4.3 TYPES OF TRAFFIC CONTROL SIGNALS

- The traffic control signals are further classified as :
  - (i)   Fixed time signals
  - (ii) Manually operated signals
  - (iii) Traffic actuated signals

(i) Fixed Time Signals :

- The signals which keeps repeating the same set of signal phases and the signal time are termed as fixed time signals.
- These signals are set to repeat a cycle of red-yellow-green lights in a fixed time at regular intervals.
- The time allocated to a traffic movement or combination of traffic movements known as phase of cycle is predetermined based on traffic studies.

  Advantages :

  - Following are the advantages of providing fixed time signals :
    - (i)   These signals are automatic and simplest.
    - (ii) These signals performs satisfactory at different approach roads with no significant variation in traffic flow.
    - (iii) Electrically operated signals are cheapest.
    - (iv) These signals are easy to use so commonly used.

  Disadvantages :

  - Following are the disadvantages of providing fixed time signals :
    - (i)   These signals may cause unnecessary delay to the traffic when traffic flow on one road is heavy as compared with other traffic flow on second road.
    - (ii) The waiting time at red phase is increased when irregular traffic flow on roads are observed.

(ii) Manually Operated Signals :

- The signals whose signal phase and signal time are operated manually are termed as manually operated signals.
- These signals are operated by traffic police constables at or near the intersection of roads.
- The signal phase and signal time varies which depends on traffic demand at that point of time.
- These signals are not common in use except cases when fixed time signals are not functioning for a while.

Distinguish between fixed time and manually operated traffic control signals :

Fixed time signals	Manually operated signals
1. These signals are operated by Power.	1. These signals are operated manually.
2. These signals are automatic.	2. These signals are not automatic.
3. The signal phase and signal time are predetermined and fixed.	3. The signal phase and signal time are varies as per traffic demand at that time.
4. These signals may cause unnecessary delay in traffic.	4. These signals avoids the unnecessary delay in traffic.
5. Waiting time at red phase increases when irregular traffic observed.	5. Waiting time can be adjusted as per traffic demand at any time during irregular traffic flow.

(iii) Traffic Actuated Signals :

- The signals whose signal phase and signal cycle can be changed according to traffic demand are termed as traffic actuated signals.
- These signals can be semi-actuated or fully actuated type.
- In semi-actuated traffic signals the green phase of an traffic approach can be extended to allow more vehicles to pass approaching closely, to clear the intersection with the help of detectors installed at approach.
- In fully actuated traffic signals, a pre-determined programming with the help of detectors at approach and traffic demand provides the right way to different traffic movements at all intersections of road.
- These signals are very costly to be installed at any intersection of road.

  Advantages :
  - Following are the advantages of providing traffic actuated signals :
    - (i) These signals reduces delaying of traffic.
    - (ii) These signals increases capacity of a road intersection.
    - (iii) These signals provide continuous traffic operation under low volume traffic.
    - (iv) These signals are adaptable to short term fluctuation in traffic.

  Disadvantages :
  - Following are the disadvantages of providing traffic actuated signals :
    - (i) Installation of detectors for these signals is costly.
    - (ii) The maintenance of detectors is difficult.
    - (iii) These signals are two to four times costlier than fixed-time signal.

(iv) The actuated controller and detectors are much more complicated than fixed-time signal.

(v) These signals are uneconomical than fixed-time signals.

4.4 LOCATION OF SIGNALS

- Following are the points to be considered while deciding location of signals :

    (i) Signals should be located at such a place that one or more faces of a signal must be visible to the drivers approaching an intersection.

    (ii) The height at which the signals are mounted on the side of a roadway varies from 2.4 to 3 m.

    (iii) In urban areas, two or more signal face installed on an approach so that drivers behind the truck get visibility of signals.

    (iv) Planning of signal position should be done according to shape of intersection of roads.

4.5 COMPUTATION OF SIGNAL TIME

- There are various methods for computation of signal time as listed below :

    (1) Fixed time cycle method

    (2) Trial cycle method

    (3) Approximate method

    (4) Webster's method

    (5) IRC method

(1) Fixed Time Cycle Method :

    (i) This is approximate methods for the design of traffic signal cycles.

    (ii) A fixed time cycle is assumed and green time periods are calculated for approach roads.

    (iii) Amber periods are assumed accordingly.

    (iv) A phase cycle length is then calculated with this data provided with fixed time cycle assumed.

(2) Trial Cycle Method :

    (i) This method is approximate method for the design of traffic signal cycles.

    (ii) During the peak hour of traffic flow the 15 counts say n_1 and n_2 on road 1 and 2 are noted.

    (iii) Some suitable trial cycle c_1 second is assumed. So the number of zoom cycles in the 15 minutes period is found to be $\dfrac{(15 \times 60)}{c_1} = \dfrac{900}{c_1}$.

    (iv) Assuming average time headway, say 25 seconds and the green periods G_1 and G_2 of road 1 and 2 respectively are calculated to clear the traffic during trail cycle –

$$G_1 = \frac{25\, n_1\, c_1}{900} \text{ and } G_2 = \frac{25\, n_1\, c_1}{900}$$

    (v) The amber periods A_1 and A_2 are calculated or rather assumed suitably.

    (vi) The length is equal to $(G_1 + G_2 + A_1 + A_2)$ seconds.

    (vii) If calculated cycle length is approximately equal to the assumed cycle length then it is accepted as design cycle length.

(viii) If not then above trial is repeated until the trial cycle length is approximately equal to calculated cycle length.

(3) Approximated Method :

(i)   This is also an approximate method as name suggested.

(ii) The following design procedure is followed for single simple design of two-phase signal unit at cross roads along with pedestrian signals.

(iii) Pedestrial signal time is calculated approximately with suitable data.

(iv) The green time along with amber time is measured with suitable data approximately.

(v)   The phase cycle length is measured with all above data approximately and suitably adopted.

(4) Webster's Method :

(i)   This is the rational method of calculation of traffic signal cylces.

(ii) As a part of field work the saturation flow 5 per unit time on each approach of water section and the normal flow q on each approach during design hour is find out.

(iii) Based on the higher value of normal flow the ratios $y_1 = \dfrac{q_1}{S_1}$ And $y_2 = \dfrac{q_2}{S_2}$ Are calculated for approach roads 1 and 2 respectively.

(iv) In case of mixed traffic, it is necessary to converted all normal flow and saturation flow values in terms of suitable PCU values which needs to be calculated separately.

(v)   During the green phases and corresponding time intervals the number of vehicles in the stream of compact flow is to be noted precisely in saturation flow.

(vi) If data is not available mentioned above then it is assumed as 160 PCU per 0.3 meter width of approach road.

(vii) The normal traffic flow on approach roads is also determined during field studies for design periods.

(viii) The optimum signal cycle C_0 corresponding to last total delayed to the vehicles is calculated as : $$C_0 = \dfrac{1.5\,L + 5}{1 - Y}$$

Where,   L = total lost time per cycle, secs = 2n + R

n = no. of phase

R = All red time

$Y = y_1 + y_2$

So, $G_1 = \dfrac{y_1\,(C_0 + L)}{Y}$ and $G_2 = \dfrac{y_2\,(C_0 + L)}{Y}$

(ix) This is adopted for greater number of signal phases.

(5) IRC Method :

•   IRC provides following guidelines for the design of traffic signal cycles :

(i)   For major and minor roads the pedestrian green time is required to be calculated at walking speed of 1.2 m/sec and initial walking time of 0.7 seconds. These are the minimum green time required for the vehicular traffic on the minor and major roads respectively.

(ii) The green time required for the vehicle traffic on the major road is increased in proportion to the traffic on the two approach roads.

(iii) The cycle time is calculated after allowing amber time of 2 seconds each.

(iv) The minimum green time required for clearing vehicles arriving during a cycle is determined for each claim of the approach road assuming that the first cycle will take 6 seconds. The subsequent vehicles of the queue will be cleared at the rate of 2 seconds. The minimum green time required for the vehicular traffic on any of the approaches is limited to 16 seconds.

(v) The optimum signal cycle is calculated as per Webster's formula. The saturation flow values may be assumed as 1850, 1890, 1950, 2250, 2550 and 2990 PCU per hour for the approach road widths 3, 3.5, 4, 4.5, 5 and 5.5 m and for widths above 5.5 m, the saturation flow may be assumed as 525 PCU per hour per m width. The lost time is calculated from amber time, inter-green time and the initial delay of 4 seconds for the first vehicle on each log.

(vi) The signal ++ cycle time and the phases may be revised keeping in view the green time required for clearing the vehicles and optimum cycle length determined in steps 4th and 5th above.

4.6 TRAFFIC ISLANDS

- The raised areas constructed within the roadway to establish physical channels through which the vehicular traffic may be guided are called traffic islands. At the traffic island all the converging vehicles are compelled to move round the island in one direction before they weave out the road, into their respective direction radiating from the island.

- Usually traffic island serves more than one function. They serve as shelter for pedestrians, while crossing the road.

- The advantages and disadvantages of providing the traffic island are discussed below:

  Advantages:

  (a) They enable all the traffic to proceed at a fairly uniform speed.

  (b) They provide an orderly organized traffic flow.

  (c) They eliminate points of direct conflict, and the accident, if occurring is of minor nature.

  (d) They provide self-governing traffic flow and need no control by police or traffic signals.

  Disadvantages:

  (a) They require comparatively larger area and may not be feasible in many built-up areas.

  (b) They make the travel troublesome, if provided at close intervals.

  (c) They result a little extra distance to be travelled by the vehicles while turning right.

  (d) They are to be supplemented by the traffic police to control the traffic in case the pedestrian traffic is more.

- The traffic islands help in controlling traffic in following manner:

  o Due to traffic islands vehicles from the converging arms are forced to move round a central island in one direction (always clock wise direction) in an orderly an organized manner and weave out of the rotary movement into their desired direction.

  o Thus, it avoids traffic congestion and provide efficient, free and rapid flow of all types of traffic on road intersection.

4.7 TYPES OF TRAFFIC ISLANDS

- The traffic islands are of following types :

  (A) Rotary or Central Islands

(B) Chanelizing or Refuge Island

(A) Rotary or central island: (W-08, W-09, S-10)

- This island is constructed in the center of an intersection to compel movement of traffic in a clock-wise direction. They eliminate points of direct conflict and provide an orderly and organized traffic flow. The main objects of providing a rotary are:

    (a) To eliminate the necessity of stopping the crossing stream of vehicles.

    (b) To reduce the area of conflict.

- In this case, all vehicles are allowed to merge into one stream around the rotary and then diverge out to the desire lane or direction radiating from the rotary. Generally, vehicles are allowed to move in clockwise direction. Thus, in this way, the crossing conflict of vehicles is avoided.

    The rotary islands are provided only when sufficient area for their construction is available at the road intersection. Their shape depends upon the type of road intersection. Generally, the following shapes are used:

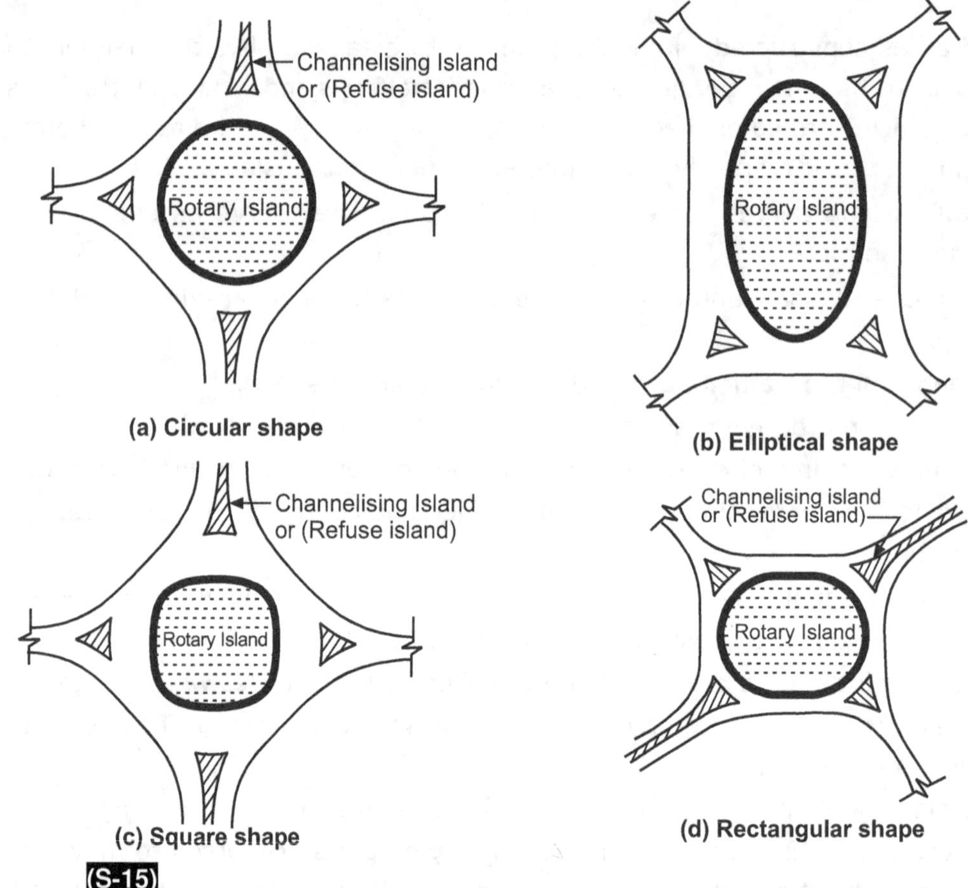

(S-15)

Fig. 4.3: Rotary Island   (W-11)

(B) Channelizing island or Refuge Island :

- These types of islands are provided to guide the traffic into proper channel through the intersection area. These islands are very useful as traffic control device for large area intersection at grade. The size and shape of these islands depends upon the layout and dimensions of intersections.

- Following are the uses of channelized islands:

    (a) They establish the desired angles of crossing and merging of traffic streams.

(b) By their introduction, the area of possible conflict reduces considerably and the vehicles can be confined to a definite path.

(c) They are useful in case, the direction of flow is desired to be changed and speed control can be forced on vehicles entering the intersections.

(d) They serve as refuged island for pedestrians and also they can be used as convenient locations for other traffic control devices.

- Typical channelized intersections are shown in Fig. 4.3, in which Fig. 4.3 (a) shows 'T' partial channelization, Fig. 4.3 (b) shows 'T' or complete channelization, Fig. 4.3 (c) shows cross or complete channelization and Fig. 4.3 (d) shows skew or partial channelization.

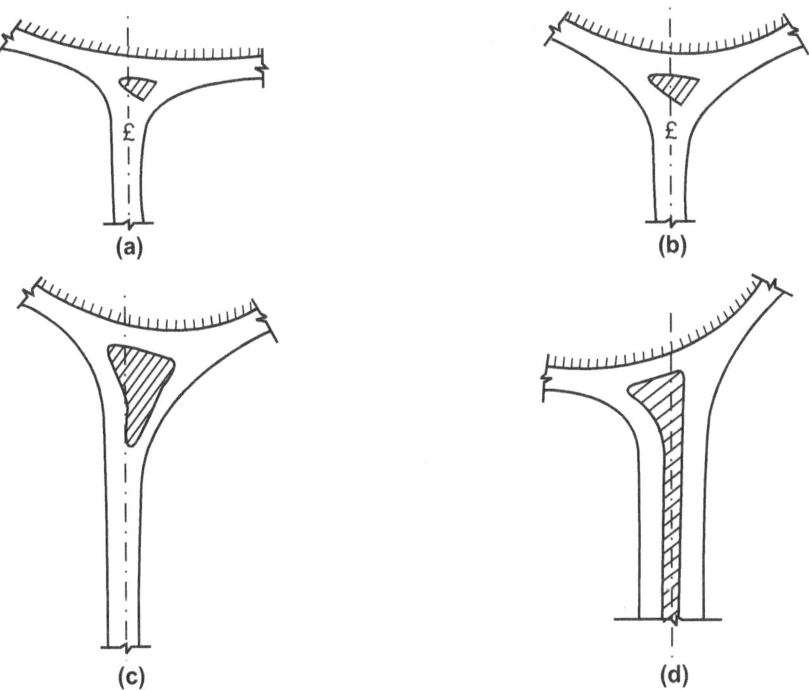

Fig. 4.3 : Channelizing Island

4.8 ROAD INTERSECTIONS OR JUNCTIONS

Definition :

- It is the place where two or more roads are arranged to join or cross at same or different levels. Actually these are the spots of accidents if proper precautions are not taken in their design or layout. Thus, safety of vehicular traffic and pedestrians is very essential at these places. Pedestrian's movements at intersections produce increased obstacles and delays.

- Therefore, at road intersections, safety of pedestrians is very essential because accident usually occur at such places. The road intersections should be well planned and signaled properly. At the time of planning and designing of road intersections following points should be kept in mind:

(a) The road intersection should as far as possible be at right angle. The acute crossing should be avoided.

(b) It should have proper provision for pedestrian crossings.

(c) It should be free from gradients and sharp curves.

(d) It should have proper drainage facilities for rain water.

(e) It should have good lighting arrangement during night hours.

(f) As far as possible, at intersections camber should be avoided and the whole area should be at same level.

- For proper visibility and future extension, sufficient space should be left.

Types :

- The road intersection may be broadly classified into the following two categories:

  (i) Intersection at-grade

  (ii) Grade-separated intersection.

4.9 INTERSECTION AT GRADE

- These are the road intersections, where all the roadways join or cross at the same level. These intersections are also called at-grade intersections. Such type of intersections are easily and economically constructed as compared to the grade-separated intersections. Following are the various types of intersections at grade:

  (a) Cross-intersection: This type of intersection at grade is provided when two roads cross each other at right angles. This is also called square intersection. Fig. 4.4 shows cross intersection.

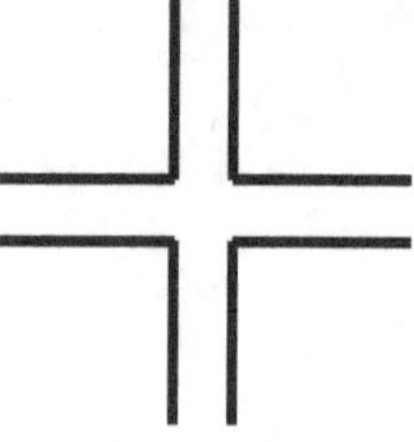

Fig. 4.4 : Cross Intersection

  (b) Tee intersection: This is the most common type of intersection at-grade. It is provided when two roads meet at each other at right angles. Fig. 4.5 shows Tee intersection.

Fig. 4.5 : Tee Intersection

  (c) Staggered intersections: When two minor roads from opposite directions meet a major road at right angles in a staggered fashion, staggered intersections are provided. In this type of intersection, minimum distance between the centerlines of minor roads meeting the major road should be 60 m. Fig. 4.6 shows staggered intersection.

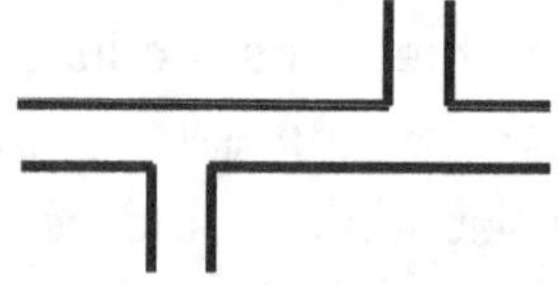

Fig. 4.6 : Staggered Intersection

  (d) Skewed intersection: When two roads meet each other at angle other than right angle, skewed intersections are provided. Fig. 4.7 shows Skewed intersection.

Fig. 4.7: Skewed intersection

(e) Cross-skewed intersection: When two roads cross each other at angle other than right angles, Cross-skewed intersections are provided. This is also known as an acute intersection or acute junction. Fig. 4.8 shows Cross-skewed intersection.

Fig. 4.8 : Cross-skewed Intersection

(f) Skewed staggered intersection: When two minor roads from opposite directions join major road at angles other than right angles in a staggered fashion, skewed staggered intersections are provided. Fig. 4.9 shows skewed staggered intersection.

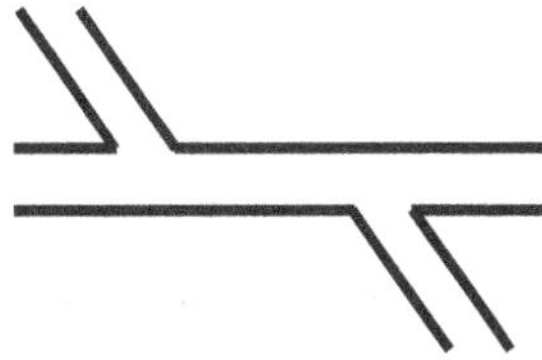

Fig. 4.9 : Skewed Staggered Intersection

(g) Wye intersection: This type of intersection at-grade is suitable when three roads join at a place at different angles. This is also known as three-way or fork intersection. This intersection proves very dangerous if not properly controlled. Fig. 4.10 shows Wye intersection.

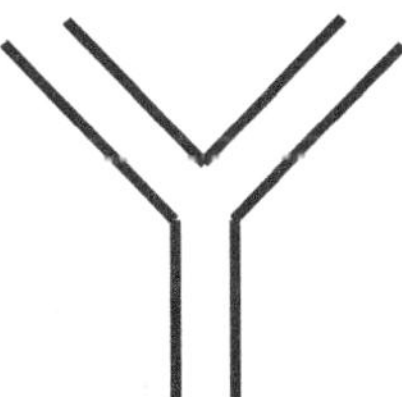

Fig. 4.10 : Wye Intersection

(h) Multiple intersections: When more than three roads join at a place at different angles multiple intersection is provided. This intersection proves very dangerous unless properly laid out and channelized. Fig. 4.11 shows Multiple intersections.

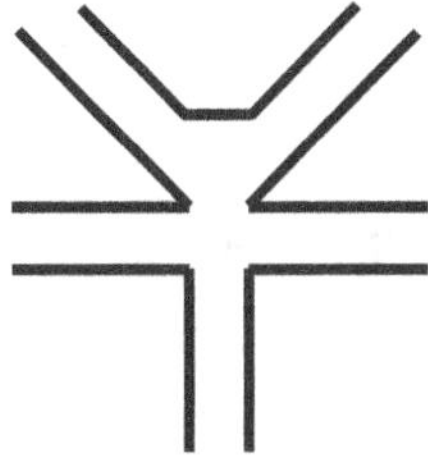

Fig. 4.11: Multiple Intersections

4.10 GRADE SEPERATED INTERSECTION

- In this type of road intersection, separation in grade of the intersecting roads is achieved by providing a bridge. The grade separation may be achieved by an over-bridge or an under pass.
- Following are the advantages and disadvantages of grade-separated intersections:

Advantages: (S-11)

(a) Grade-separated intersections provide increased safety for turning traffic and by introducing indirect interchange ramp, even right turn movements can be made quite easy and safe.

(b) Grade-separated intersections provide maximum facility to the crossing traffic and avoid accident while crossing.

(c) Grade-separated intersections provide an overall comfort and convenience to the motorists and saving travel time.

Disadvantages: (S-11)

(a) Grade-separated intersections are very costly in their construction in order to obtain complete grade separation and interchanged facilities.

(b) Grade-separated intersections may cause undesirable crests and sag in vertical alignment in flat or plain areas.

(c) Grade-separated intersections are difficult and undesirable where there is a limited right of way like built-up or urban area or where the topography is not favourable.

- Transfer of route at grade-separated intersection is provided by interchanged facilities of following type of ramps:

1. **Direct ramp:** The direct interchanged ramp involves diverging to the right side and merging in from the right. Fig. 4.12 shows direct ramp.

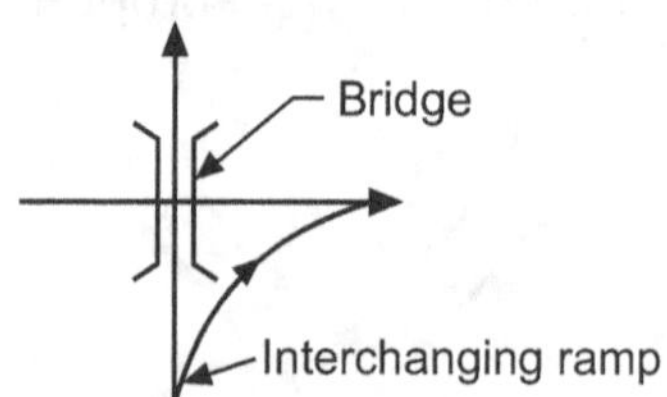

Fig. 4.12: Direct Ramp

2. **Semi-direct ramp:** The semi-direct interchanged ramp allows diverging to left but merging in from right side. Fig. 4.13 shows Semi-direct ramp.

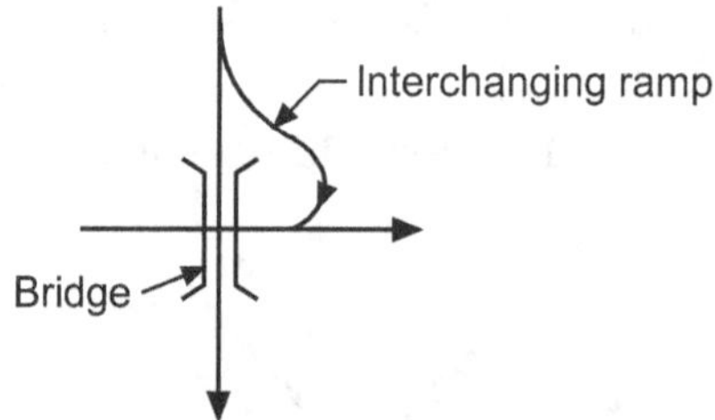

Fig. 4.13 : Semi-direct Ramp

3. **Indirect ramp:** The indirect interchange ramp is a simple diverging to the left and merging in from the left side and thus the distance travelled in this case is minor. Fig. 4.14 shows Indirect ramp.

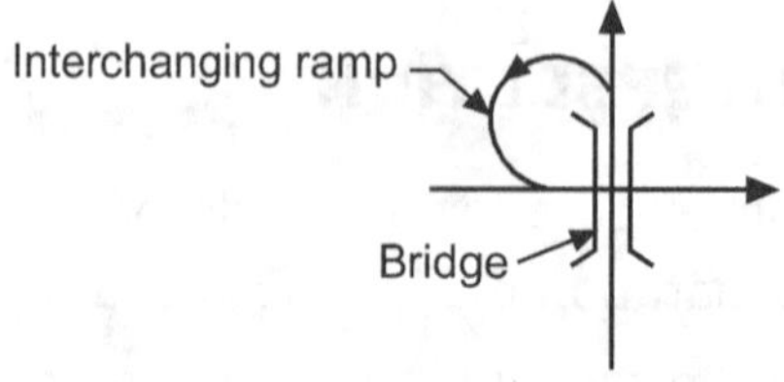

Fig. 4.14 : Indirect ramp

- The Grade-separated intersections are classified into the following two types:

  (a) Over passes.

  (b) Under passes.

(a) Over Passes:

- These are the grade-separated intersections in which major highway is taken above in embankment across the other highway by constructing an over bridge. This is also called *fly-over*.
- Following are the advantages and disadvantages of over passes or fly-over.

  Advantages:

  (a) An over pass or fly-over reduces the cost of the bridge, being small in span over the crossroad.

  (b) An over pass or fly-over reduces the troublesome drainage problems.

  (c) An over pass or fly-over enables the future expansion of both the roads and thus stage construction of the bridge structure is possible.

  (d) An over pass or fly-over provides an aesthetic preference to the main through traffic and less feeling of restriction on the major highway.

  Disadvantages:

  (a) An over pass or fly-over provides restriction to sight distance unless long vertical curves are provided.

  (b) An over pass or fly-over may cause speed reduction of heavy vehicle moving on the major highway, having steep gradient in rolling terrain.

  Following are the types of over pass or fly-over:

  (i) Clover-leaf junction (S-14): This type of grade-separated intersection arranged for two-road crossing at right angles so that the vehicle can keep to the left only, all along the process of turning towards the road on its right. The provision of this type of grade-separated intersection pattern is desirable when the traffic volume at a road intersection is above 5,000 vehicles per hour. This type of grade-separated intersection fulfills all the requirements of turning traffic involving the simplest traffic manoeures, viz., diverging to the left and merging from left by providing four indirect interchanged ramps. Fig. 4.15 shows a clover leaf junction type fly-over or an over pass.

(S-14)

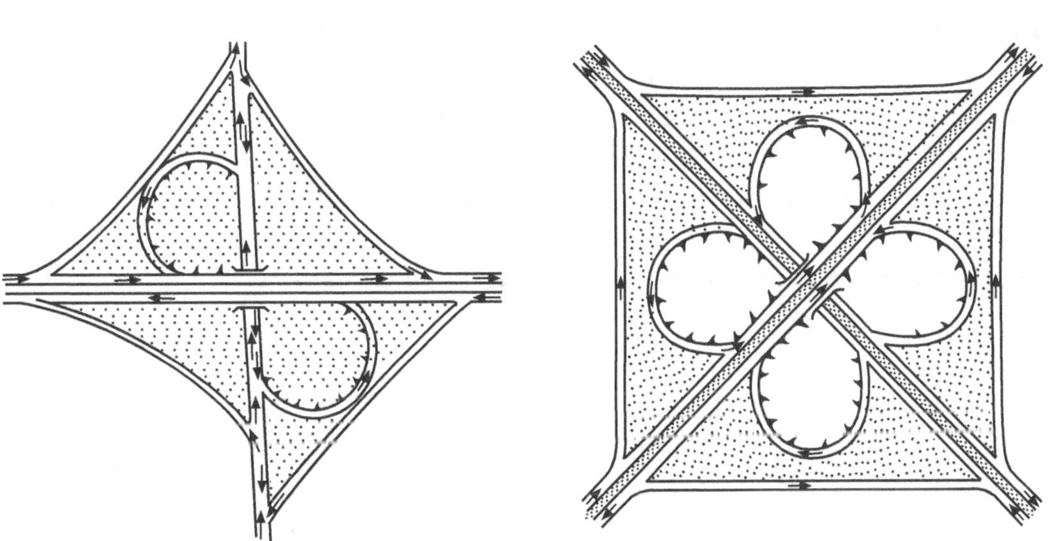

(a) Partial Clover Leaf   **(b) Full Clover Leaf**

Fig. 4.15: A Clover Leaf Junction Type Fly-over or An Over Pass

(ii)Trumpet type fly-over: This type of grade-separated intersection is arranged for two roads meeting at right angles so that the vehicle can keep to the left only, all along the process of turning towards the road on its right. Fig. 4.16 shows trumpet type fly-over.

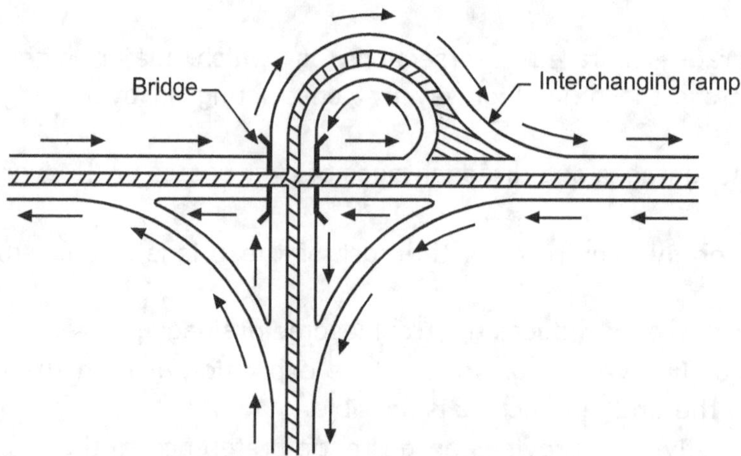

Fig. 4.16: Trumpet Type Fly-Over

(b) Under Passes:

- These are the grade-separated intersections in which one highway is taken by depressing it below ground level across the highway by constructing an under bridge. This type of grade-separated intersection is taken along the existing grade without alteration of its vertical alignment and the cross road is depressed and taken underneath by constructing an under bridge.

- Following are the advantages and disadvantages of providing under passes:

Advantages:

(a) An under pass gives warning to the traffic in advance due to the presence of an under bridge which can be seen from a distance.

(b) If major highway is taken below, the under pass is advantageous to the turning traffic because the traffic from the cross road can accelerate while descending the ramp to the major highway and the traffic from the major highway can deaccelerate while ascending the ramp to the cross road.

Disadvantages:

(a) There is feeling of restriction to the traffic while moving along the under pass.

(b) An under pass may create troublesome drainage problems, during rainy season when the ground water table rises high.

(c) The overhead structure may restrict the vertical sight distance even at the valley curve near the under pass.

(d) No possibility of future expansion of roads at the underpass.

4.11 SEGREGATION OF TRAFFIC

Definition :

- The techniques used to separate the traffic for safe and easy movement of vehicles and/or pedestrians at any particular time and place is called as segregation of traffic.

Purpose :

- The following are the purposes behind providing segregation of traffic :

(i) To increase the capacity of existing roads.

(ii) To regulate the traffic at intersections.

(iii)To increase the safety of vehicles and/or pedestrians.

(iv)To understand and to improve traffic situation.

(v) To reduce traffic congestion at road intersections.

(vi)To achieve better utilisation of resources.

Types :

- The types of segregation of traffic are as follows :

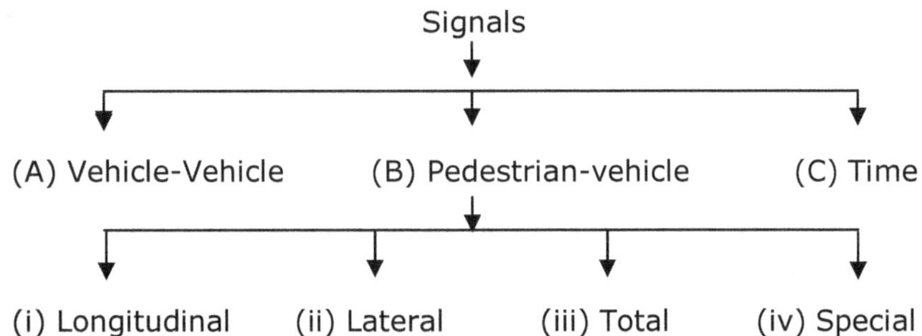

(A) Vehicle-vehicle Segregation :

- In this type of segregation of traffic vehicles are separated from another vehicles.
- In this fast moving vehicles are separated from slow moving vehicles.
- It is observed in areas where slow moving vehicles are confined to outer areas.

(B) Pedestrian-vehicle Segregation :

- In this type of segregation of traffic pedestrians are separated from another vehicles.
- This type of segregation of traffic can be achieved by following sub-types :

  (i) Longitudinal Segregation :

  o In this type of pedestrian-vehicle segregation the separators are provided along the length of road and are parallel to the roads.

  o The footpath provided for pedestrians is example of longitudinal segregation of pedestrians and vehicles.

  (ii) Lateral Segregation :

  o In this type of pedestrian-vehicle segregation the separators are provided across the length of road and are perpendicular to the roads.

  o The zebra crossings, foot over bridges, pedestrian sub-ways, pedestrian traffic signals are some examples of lateral segregation of pedestrians and vehicles.

  (iii) Total Segregation :

  o In this type of pedestrian-vehicle segregation the separators are provided such that pedestrians are separated from moving vehicles totally.

  o This is done in case of the large movement of pedestrians.

  o In such cases separate streets are provided for pedestrians.

  o Pedestrian mall is an example of total segregation of pedestrians and vehicles.

  (iv) Special Segregation :

  o This is the segregation provided specially for the cyclist for their safe and easy movement from vehicles.

(C) Time Segregation :

- In this type of segregation of traffic regulatory time is allotted to certain vehicles in a day.
- Specific traffic is regulated by giving specific time period during day.
- For an example, heavy vehicles are not allowed during peak hours in busy areas of crowded cities.

Important Points

- The devices used for controlling, warning or guiding the traffic are defined as traffic signals.
- Types of Signals :
  - (A) Traffic Control Signals
  - (B) Pedestrian Signals
  - (C) Special type traffic Signals
- There are various methods for computation of signal time as listed below :
  - (1) Fixed time cycle method
  - (2) Trial cycle method
  - (3) Approximate method
  - (4) Webster's method
  - (5) IRC method
- The traffic islands are of following types :
  - (1) Rotary or Central Islands
  - (2) Chanelizing or Refuge Island
- These are the road intersections, where all the roadways join or cross at the same level. These intersections are also called at-grade intersections.
- These are the grade-separated intersections in which major highway is taken above in embankment across the other highway by constructing an over bridge. This is also called fly-over.
- The techniques used to separate the traffic for safe and easy movement of vehicles and/or pedestrians at any particular time and place is called as segregation of traffic.

Practice Questions

1. Define following terms :
   - (i)   Traffic Signals,   (ii) Road Junctions,
   - (iii) Grade intersection,   (iv) Fly-over
   - (v)   Segregation of traffic
2. State and Explain types of signals.
3. Enlist advantages of Traffic Control Signals.
4. What are the types of traffic Control Signal. Explain any one in brief.
5. State and explain methods for computation of signal time.
6. Write short note on : (i) Approximated Method, (ii) Webster's Method.
7. What is Traffic Islands ? Also state advantages and disadvantages of it.
8. State the types of Traffic Islands. Explain any one in brief.
9. Write note on clover calf junction.
10. What is time Segregation ?

☆☆☆

Chapter 5

ROAD ENVIRONMENT AND ARBORICULTURE

Syllabus

5.1 **Street lighting :** Definition, sources necessity, types – luminaire, foot candle, lumen, factors affecting their utilization and maintenance.

5.2 Factors affecting visibility at night.

5.3 **Arboriculture :** definition, objectives, factors affecting selection of type of tree.

5.4 **Maintenance of trees :** protection and care of road side trees

Objectives

5a Suggest the street lighting system for the given road section.

5b. Recommend the type of tree for road side plantation.

5c. Justify the need of protecting the road side plantation.

5d. Describe the methods of protecting the road side plantation.

5.1 STREET LIGHTING

5.1.1 Defination

- It is a system which gives source of light to the road users to safe, comfortable and convenient use of road.

- It is useful to see accurately and easily the road/carriage way and immediate surrounding in darkness.

5.1.2 Necessity of Street Lighting

(1) It provides safety to the road users.

(2) It provides security on the road.

(3) It helps to reduce road accidents.

(4) It reduces street crime.

(5) It reduces glare from vehicle headlights.

(6) It helps for easy movement of vehicles.

(7) It helps for easy driving at intersections.

(8) It helps for easy movement of traffic in bridge, flyover etc.

5.1.3 Sources of Street Lighting

(A) Following are the sources of street lighting :

    (a) Filament

    (b) Mercury vapour

    (c) Fluorescent

    (d) Sodium vapour

    (e) Solar

(B)

    (a) Tungesten Filament

    (b) Tubular Fluorescent

    (c) Sodium Vapour Discharge

    (d) High pressure mercury Discharge

    (e) Solar System

5.1.4 Types of Street Lighting

(A)

    (1) Tungsten filament

    (2) Fluroescent tubular

    (3) Low pressure sodium

    (4) High Pressure Sodium (HPS)

    (5) Mercury bulb (MB)

    (6) Mercury bulb (Fluorescent) (MBF)

    (7) LED solar street light (Light Emitting Diode)

    (8) Street light solar

    (9) High efficiency LED

    (10) Narrow – Band Amber (NBA) LED street light

    (11) Phosphor – Converted Amber (PCA) LED street light

(B)

    (1) High iumen street light : 60 Watt

    (2) Street light solar (30 W to 300 W)

5.1.5 Important Definations

(1) Luminous Flux (F) :

- It is the radient power given by light source. It is denoted by F.

(2) Lumen (lm) :

- It is the unit of luminous flux.

(3) Lux :

- The lux (lx) is the SI derived unit of illuminance and luminous emittance, measuring luminous flux per unit area.

- It is equal to one lumen per square meter.

(4) Candela :

- It is the base unit of luminous intensity in SI.
- It is luminous power per unit solid angle emitted by a point light source in a particular direction.

(5) Luminous intensity (I) :

- It is an expression of the amount of light power emanating from a point source within a solid angle of one steradian.

(6) Meter Candle :

- A unit of illuminating power, the light given by one standard candle at a distance of 1 m.

(7) Meter Candle Second :

- A unit of exposure of light equivalent to an illumination of one lux for one second.

(8) A foot candle :

- A foot candle is a non SI unit of illuminance intensity.
- One foot candle represents 'the illuminance cast on a surface by a one-candela source one foot away 1 FC ≈ 10.76 lux'.

(9) Illumination (E) :

- When a light falls on a surface, it becomes visible, the phenomenon is called as illumination.
- It is defined as luminous flux falling on a surface per unit area.
- It is measured in lumen per square meter or meter-candle.

5.1.6 Factors Affecting Utilization and Maintenance of Street Lighting

- The factors which affects the utilization and maintenance of street lighting are given below :

(1) Source/Types of Light :

- When source/types of light is main factor as per as utilization and maintenance is concern.
- Now-a-days LED's arc preferred. They gives pleasant light in minimum Watts. There maintenance is also low. But their initial cost is high.
- Now-a-days solar light system is also provided. Solar light system is eco friendly.

(2) Luminaire Distribution of Light :

- Luminaire distribution of light is one of the important factor as it is related to utilization of light.
- Proper Luminaire distribution covers the required area on road and it gives uniform and bright light on pavement.
- Proper Luminaire distribution of light cover the area between pavement and Kerb. It also helps to see road signs and markings.

(3) Spacing of Lamp (lanterns) :

- For better visibility and brightness on the road closer spacing of lamp is required.
- As per requirement and budget the different arrangements of lanterns (lamp) are made. These arrangements are :
  - (i) Staggered
  - (ii) Opposite

(iii) Central

(iv) Single sided

(4) Mounting height :

- It is the vertical distance between the center of the lamp (lantern) and the carriage way.
- This height is important for proper distribution of light.
- Shadow effect, glare effect are depends upon this height.
- Mounting height at lamp varies from 6 m to 9 m.

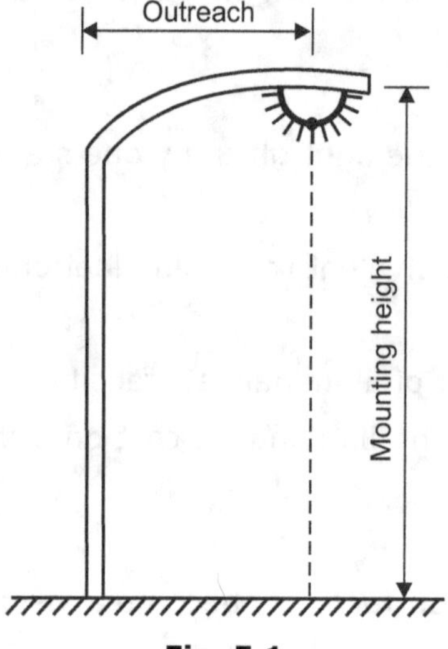

Fig. 5.1

(5) Overhang (outreach) of lamp :

- This is the distance measured horizontally between the centre of lamp (lantern) and the centre of the column or pole.
- This factor is responsible for proper distribution of light, area covered by the light uniform, brightness of the light on road.

(6) Types of Pavement :

- It is observed that light coloured surfaces (Cement Concert Road) gives good reflection of light as compare to dark colour surfaces (flexible pavement).
- Rigid pavements (Cement Concrete Road) are more visible than flexible pavement, during night time.
- Rough surfaces are not desirable for proper distribution and effects of light.

(7) Prevention of Glare :

- Glare effect is essential to reduce for safety purpose.
- Glare effect can be decreased by providing high mounting of lamp.
- We can reduce glare effect by decreasing luminaire brightness by increasing the area.

(8) Planning Road Side Development :

- Proper planning and development of highway/express way affects the lighting system.
- Proper planning make it more effective, economical and safe.

5.2 FACTORS AFFECTING VISIBILITY AT NIGHT

1. Amount of Light

2. Distribution of Light

3. Size of object

4. Brightness of object

5. Types of lamps

6. Glare on the driver's eye

7. Mounting height

8. Brightness of background

9. Colour or pavement

10. Reflecting characteristics of pavement surface

11. Time available to see the object.

SOLVED EXAMPLE

Example 5.1 : *Calculate the spacing between lighting units for design of street lighting system. Street width = 16 m, Mounting height = 8 m, Lamp size = 6000 lumens, Flux = 6.0 lux, Coefficient of utilization = 0.36, Maintenance factor = 0.8.*

Solution :

(1) $\text{Ratio} = \dfrac{\text{Street widht}}{\text{Mounting height}} = \dfrac{16}{8} = 2$

(2) Spacing between lamps

$$= \frac{\text{Lamp lumen} \times \text{Coeff. of utilization} \times \text{Maintenance factor}}{\text{Average flux} \times \text{Width of load}}$$

$$= \frac{6000 \times 0.36 \times 0.8}{6 \times 16} = 18 \text{ m}$$

∴ Adop spacing of 18 m between lights.

5.3 ARBORICULTURE

5.3.1 Defination of Arboriculutre

- Arboriculture is defined as planning of plantation of trees along the road land.

- This is important aspects of road side development.

- It give attractive, pleasant landscape to the road sides.

5.3.2 Purpose/objective/ use of Arboriculture

1. It provides shades to the road users.

2. It gives attractive landscape.

3. It gives pleasant drive to the road users.

4. It protects environmental degradation.

5. It increases level of oxygen.

6. It reduces the sound.

7. It intercept fumes from road vehicles.

8. It protects the soil erosion along the road side.

5.3.3 Factors Affecting Selection of Type of Tree

(1) Crown of the tree :

- Wide crown trees are not preferred, as they obstruct day light.
- It is noted that crowns of the tree planted on both sides of the road should not cover carriage way.
- It is also noted that, crown of trees do not extend beyond the pavement edges.

(2) Soil type : Plantation is depends upon the type of soil. For clayey soil Jamun, Mango trees are preferred.

- In sandy soil : Shisham, Kaju and Arroo are suitable.
- Loamy soils are suitable for plantation of all types of trees species.
- In case of water logged areas, Jamun, Arjun, Tarcharbi are suitable.

(3) Climate : Climate plays important role in the growth of trees. Selection of trees should be inconsideration with climate. In dry and hot climate trees like Neem can survive easily.

5.4 MAINTENANCE OF TREES PROTECTION AND CARE OF ROAD SIDE TREES

5.4.1 Location of Tree Planting

- It is required to decide the location of trees along the road side.
- Their spacing is marked and pits for planting the trees are then prepared at the selected locations.
- Plants are selected from nursery having size 1 to 1.5 m tall and the steam of diameter is about 20 mm.
- Sufficient manure is required to placed at the bottom of the pit.

5.4.2 Protection of Care of Trees

- After planting the trees, for their proper growth protection is needed.
- For proper protection,
  - (1) Steel Guard
  - (2) RCC Guard
  - (3) Brick Work Guard
  - (4) Plastic Drums
  - (5) Trenches are used to protect the trees.
- Generally trenches are more effective in rural areas.

5.4.3 Amount of Watering to the Trees

- Amount of watering depends upon :
  - (i) Type of soil
  - (ii) Type of plant
  - (iii) Climatic condition
- Proper watering is needed to do for atleast fixed two to three years.
- After this most of the trees are well established.

5.4.4 Care of Trees

- For proper development and well being of trees through their life span following operations are required to do.

  (i) Pruning

  (ii) Lopping

  (iii) Felling

(i) Pruning :

  - Pruning is done to maintain the symmetry of tree.
  - In this unwanted growth of branches are removed.
  - Pruning help them to grow straight and be more.
  - Generally cold weather season is chosen for this operation.
  - Sharp knife or a special pair of seissors is used to cut the unwanted growth of branches.
  - After this wounds of the trees are covered with bitumen coating.

(ii) Lopping :

  - For the roadside trees, certain shape of trees is required to maintain.
  - To train the trees to develop a particular shape during its growth, Lopping process is done.
  - In this extra branches of trees are removed, which otherwise disturb the traffic.
  - Generally in the month of February or in the month of September lopping is carried out.

(iii) Felling :

  - Felling is the process of removing or cutting the trees. To remove over-matured trees it is done. It is also done when the trees are too closely spaced.
  - In this process, earth around the trees is removed, upper roots are exposed and then trees are removed.

5.4.5 Name of the Trees Planted on Road Sides

(1)	Mango	(2)	Neem	(3)	Jamun	(4)	Amltas
(5)	Gulmohar	(6)	Ashok	(7)	Arjun	(8)	Arroo
(9)	Banyan	(10)	Kanji	(11)	Shisham	(12)	Tarcharbi
(13)	Kaju	(14)	Tun	(15)	Imli	(16)	Bargad
(17)	Kachnar	(18)	Amaltas				

Important Points

- It is a system which gives source of light to the road users to safe, comfortable and convenient use of road.
- The factors which affects the utilization and maintenance of street lighting are given below :
  - (a) Source/Types of Light
  - (b) Luminaire Distribution of Light
  - (c) Spacing of Lamp (lanterns)
  - (d) Mounting height

- (e)   Overhang (outreach) of lamp
- (f)   Types of Pavement
- (g)   Prevention of Glare
- (h)   Planning Road Side Development
- Arboriculture is defined as planning of plantation of trees along the road land.
- Amount of watering depends upon :
- (i)   Type of soil
- (ii)   Type of plant
- (iii)   Climatic condition

Practice Questions

1. Define street lighting.

2. Give any four necessity of street lighting.

3. Enlist sources of street lighting.

4. Enlist types of street lighting.

5. Define

   (i)   Lumen

   (ii)   Luminous Intensity

   (iii) Candela

   (iv) Illumination

6. Explain factors effecting utilization and maintenance of street lighting.

7. Enlist any eight factors affecting visibility at night.

8. Define : Arboriculture

9. Give purpose of Arboriculture.

10. Explain factors affecting selection of type of trees.

11. Write a note on maintenance of trees.

12. Name any four trees planted on road side.

☆☆☆

ROAD ACCIDENT STUDIES

Objectives

6a. Analyze the causes of accident occurred across the given section of road.

6b. Suggest preventive measures to avoid the accidents on a road.

6c. Create awareness about the traffic rules and laws to be followed for safe traffic movement.

6.1 INTRODUCTION

- The increase in road traffic has been followed by steep increase in the road accident. It is estimated that nearly 60,000 are killed on the roads every year. The problem of accident is very acute in highway transportation due to complex flow patterns of vehicular traffic, presence of mixed traffic and pedestrians.

- Traffic accident may involve property damages, personal injuries or even casualties. One of the main objectives of traffic engineering is to provide safe traffic movements.

- Road accident cannot be totally prevented, but by suitable traffic engineering and management measures, the accident rate can be decreased considerably. Therefore the traffic engineer has to carry out systematic accident studies to investigate the causes of accident and to take preventive measures in terms of design and control.

6.1.1 The Various Objectives of Accident Studies

- The various objectives of accident studies are:

   (a) To study the causes of accident and to suggest remedial measures.

   (b) To check the existing designs and put the suggestions for the proposed design.

   (c) To carry out before and after studies and to demonstrate the improvement in the problem.

   (d) To make computation of financial loss and to justify the proposal for improvement.

   (e) To give economic justification for the improvements suggested by the traffic engineer.

6.1.2 Classification of Road Accidents 

- The road accidents may be grouped in the following two groups:
    1. **Collision accidents:** Under this head, following types of accidents may be grouped:
        (i)   Collision with pedestrians.
        (ii) Collision with other motor vehicles, train, cycle and any other objects etc.
        (iii) Collision with fixed objects (i.e. building, kerbs, traffic separators etc.).
    2. **Non-collision accidents:** Under this head, following types of accidents may be grouped:
        (i)   Running-off the roadway.
        (ii) Over turning
        (iii) Any other non-collision on roadway.

6.1.3 Necessity of Accident Studies

(1) To study the causes of accidents.
(2) To suggest engineering measures to reduce the accidents.
(3) Measurement of length of skid.
(4) Recording the relative positions of vehicles and objects involved in accidents.
(5) To study condition of pavements.
(6) To study defects in pavements.
(7) Environmental studies and weather conditions responsible for accidents.
(8) Preparation of collision diagram.
(9) Preparation of condition diagram.
(10) To prepare accident investigation report.

6.1.4 Collision Accidents

- A collision accidents is also called motor vehicle collision (MVC).
- This type of accidents occurs when a vehicle collides with other vehicle, or pedestrian or animal or any other stationary objects such as tree, pole or building.

Type of Collision Accidents :

(1) Head on collision
(2) Right angled collision
(3) Right turn collision
(4) Rear end collision
(5) Side impact collision

6.1.5 Non-Collision Accidents

- A motor vehicle accidents which does not involve a collision is called as non-collision accidents.
- Non-collision accidents are :
    (i)   Jackknifes (Folding of vehicle)
    (ii) Overturns
    (iii) Fires
    (iv) Cargo shifts and spills
    (v)   Run-off the road

(vi) Rollover accidents

6.2 CAUSES OF ACCIDENTS

- There are four basic elements in a traffic accident i.e. the road users, the vehicle, the road and its condition (traffic) and environmental factors (weather) etc. The road users responsible for the accident may be the driver of one or more vehicles involved, pedestrians or the passengers. Vehicles involved in the accident may also be defective.

- The condition of the road surface or other existing geometric features or any of the environmental conditions of the road may not be upto the expectation causing an accident. The various causes of the road accident may hence be listed as given below.

    (a) Drivers: Use of excessive speed by the driver, rash driving, carelessness, violation of rules and regulations, failure to see or understand the traffic situation, sign or signal, temporary effects due to fatigue, sleep alcohol.

    (b) Pedestrians: Violation of traffic riles, wrong use of carriageway etc. Similarly passengers at the time of alighting and getting in do not follow traffic rules.

    (c) Element of Surprise: Change in single timing without prior information etc.

    (d) Passenger: Alighting from or getting into moving vehicles.

    (e) Vehicle defects: Failure of brakes, steering system, or lighting system, tyre burst any other defect in the vehicles.

    (f) Road condition: Slippery or skidding road surface, potholes, ruts and other damaged conditions of the road surface.

    (g) Road design: Defective geometric design like inadequate sight distance, inadequate width of shoulder, improper curve design, improper lighting, and improper traffic control devices.

    (h) Weather: Unfavorable weather conditions like mist, fog, snow, dust smoke or heavy rainfall, which restrict normal visibility and render driving unsafe.

    (i) Animals: Stray animals on the road.

    (j) Other causes: Incorrect signs or signals, gate of level crossing not closed when required, ribbon development, badly located advertisement boards or service station etc.

    (k) Lack of prior lighting etc.

6.3 PREVENTIVE MEASURES

- The various measures adopted to prevent the accidents may be sub-divided into the following three main groups:

    (a) Engineering measures

    (b) Enforcement of regulations

    (c) Education of public in traffic laws.

    (a) Engineering Measures: These preventive measures further can be sub-divided as follows:

    (i) Road design: The geometric features of roadway such as width of pavement, horizontal and vertical alignments, grades, sight distance etc. should be checked and redesign if necessary. The characteristics of the busy or congested route should be maintained of the desired standard. To minimize the conflict points, separation structure such as over bridge and under pass should be provided.

(ii) Road lighting: Poor lighting is a source of an accident. Hence, proper lighting on roads can minimize the rate of accidents especially during the night time. Lighting is particularly at intersections, bridges and at the places where there are restrictions to traffic movements.

(iii) Maintenance of defective vehicles: The most essential system of vehicle such as breaking system, steering and lighting system etc. should be checked frequently and defect observed should be rectified without delay.

- Engineering measures taken to reduce the accidents have proved very encouraging. The results of a study are shown in Table.

Table 6.1 : Engineering measures taken to reduce the accidents

Proper super-elevation of an isolated bend rural road.	Reduction in 60% personal injury accidents.
Improvement in visibility at junctions of rural roads.	0% reduction of personal injury accidents.
Staggering of cross road in rural areas.	60-85% reduction of personal injury accidents.
Removal of obstructions at edge of carriageway and provision of safe guards.	0-12% reduction in all types of accidents.
Improvement of bends on rural road by realignment.	80% reduction in personal injury accidents.
Improvement of sign and guidance.	Upto 25% reduction of accident at the location.
Provision of proper signs and road markings at junctions in rural areas.	Upto 50% reduction in personal injury accidents.
Channelization of intersections.	Upto 20% reduction of accidents at the junctions.
Provision of acceleration and deceleration lanes at junctions.	Upto 10% reduction in accidents at the junction.

(b) Enforcement of Regulations: These preventive measures further can be sub-divided as follows:

(i) Speed control: Generally drivers of public carrier etc. are in habit of fast driving. In order to develop the habit in drivers to drive at permissible speed, speed recorders such as tachometers should be fitted in the vehicles. The tachometer can locate the distance in which the driver has gone over speed. Secondly, surprise checks on spot speed of all fast moving vehicles may be done at selected locations and timings. The strict action should be taken against those who violate the speed limit and other regulations.

(ii) Traffic control: The signal system should be maintained in proper working conditions. If at some location it is not there, it should be introduced. Secondly proper traffic control devices such as sign, markings etc. should be installed.

(iii) Training and supervision: Strict supervision and testing should be done at the time of issuing driving licenses to the driver. The trained driver should be tested under proper supervision periodically and they should also train in proper defensive driving.

(iv) Observance of law and regulation: This is one of the most important steps in enforcement for prevention of accidents. The traffic authorities should send a team of trained personnel assisted by police to different locations to check if the road users are following the correct road regulations. This study will provide a useful data for future guidance in revising the traffic regulations.

(v) Medical check-up: The drivers are in habit of consuming alcohol or other intoxicating drinks. Therefore, they should be checked before they take out their vehicles from workshop or depot. Public carrier driver should also be checked up for vision and reaction time periodically.

(c) Education of Public in Traffic Laws: It is very essential to educate the road users regarding necessary safety precautions taken while using the road. They should be taught how to get down or get in the vehicles etc.

6.4 ACCIDENT REPORT

- After the incident of accident, the accident should be reported to nearby police station/police authorities.
- Police authorities further collect required details of accident.
- Police authorities take legal action as per the condition of accidents i.e. injuries, casualties or severe damage to property.
- After collecting accident data accident report is prepared with all facts.
- This report might be useful for subsequent analysis, evaluation of accident cost, further legal procedure, claims for compensation, insurance etc.

6.4.1 Accident Records

- Accidents records are maintained with all details related to accidents.
- It contains location of accidents, time of accidents and other details related to accidents.
- The content of accident records are :
  (i)   Location files
  (ii) Spot maps
  (iii) Collision diagram
  (iv) Condition diagram

  (i) Location Files :
  - These files are useful to keep record of location of accidents. It indicate the zone of the area. It helps to identify location of high accident incidence.
  - Location files are maintained by each police station for respective legal action and jurisdication.

  (ii) Location Spot Files : Location spot maps show accident by spots :
  - This maps are drawn by a scale 1 mm = 1 to 6 m.
  - It is used for making spot maps of urban accidents.
  - It shows total and non-total accidents by symbols.

Types of accidents	Fatal	Non-Fatal
(A) Motor vehicle pedestrian	◉	⊙
(B) Other Motor vehicle traffic	●	●

6.5 COLLISION DIAGRAM

- Collision diagram give the judgement of accident location.
- It also show the approximate path of the vehicles and pedestrians involve in the accidents.
- It also show the path of vehicles and other objects with which the vehicles have collided.
  Use of Collision Diagram:
  1. Collision diagram are most useful to compare the accident pattern before and after the remedial measures have been taken.

2. Collision diagram are used to display and identify similar accident patterns.

3. It also gives the information of type and number of accidents.

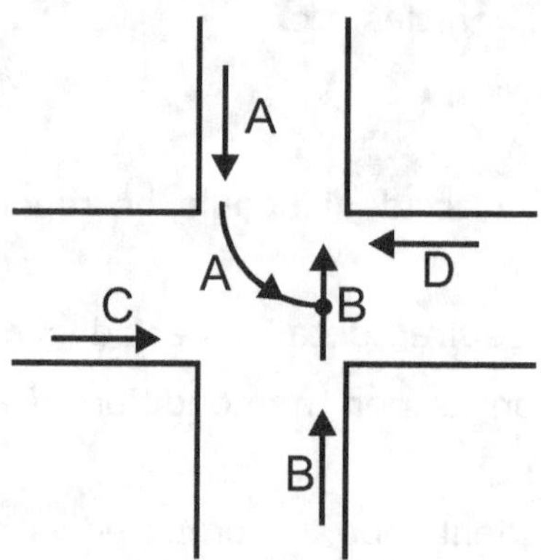

Fig. 6.1 (a) : Collision Diagram

4. Collision diagram gives the information about other factors related to accidents like:

   (a) Time of day

   (b) Day of week

   (c) Climatic conditions

   (d) Pavement conditions

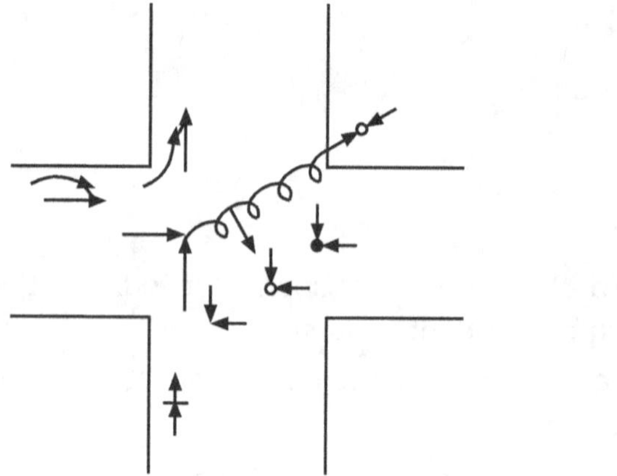

1)   Parked vehicle

2)   Side swipe

3)   Personal injury

4)   Fatal accidents

5)   Property damage only

6)   Fixed object

7)   Motor vehicle warning ahead

8)   Rear and collision

Fig. 6.1 (b): Collision diagram with symbol

6.5.1 Condition Diagram

- A condition diagram is a drawing of the accident location drawn to scale.

- Scale 1 : 100 to 1 : 250 is used for drawing of condition diagram.

- Condition diagram shows all important physical features of the road and adjoining area.

- The important features to be shown in condition diagram are :

  - Width of roadway
  - Shoulders
  - Curves
  - Kerblines
  - Bridges/overbridges/underpass
  - Culverts
  - Electric posts
  - Trees
  - Property lines
  - Footways
  - Driverways
  - Cycle track, if any
  - Sight obstruction in Roadways
  - Traffic signs
  - Signals
  - Markings
  - Street lighting

- To prepare condition diagram standard symbols are used.
- It required, the condition and collision diagram may be combine together.

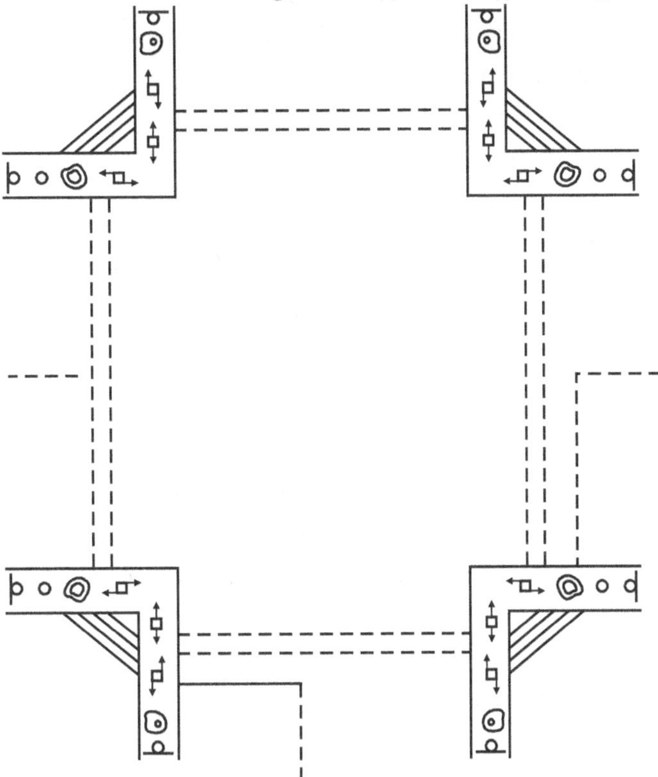

Fig. 6.2 : Typical Condition Diagram

Symbols for condition diagram

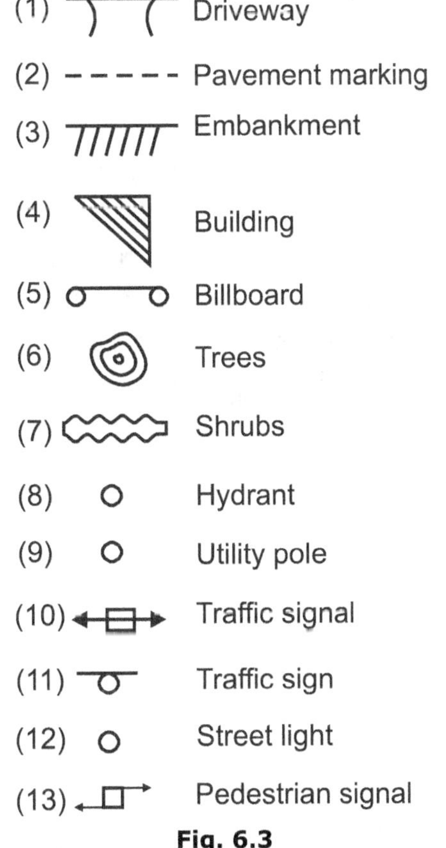

(1)   Driveway

(2)   Pavement marking

(3)   Embankment

(4)   Building

(5)   Billboard

(6)   Trees

(7)   Shrubs

(8)   Hydrant

(9)   Utility pole

(10)   Traffic signal

(11)   Traffic sign

(12)   Street light

(13)   Pedestrian signal

Fig. 6.3

Important Points

- Traffic accident may involve property damages, personal injuries or even casualties. One of the main objectives of traffic engineering is to provide safe traffic movements.
- Road accident cannot be totally prevented, but by suitable traffic engineering and management measures, the accident rate can be decreased considerably.
- The road accidents may be grouped in the following two groups: Collision accidents and Non-collision accidents.
- Type of Collision Accidents :
  - (i) Head on collision
  - (ii) Right angled collision
  - (iii) Right turn collision
  - (iv) Rear end collision
  - (v) Side impact collision
- Non-collision accidents are :
  - (a) Jackknifes (Folding of vehicle)
  - (b) Overturns
  - (c) Fires
  - (d) Cargo shifts and spills
  - (e) Run-off the road
  - (f) Rollover accidents
- The various measures adopted to prevent the accidents may be sub-divided into the following three main groups:
  - (a) Engineering measures
  - (b) Enforcement of regulations
  - (c) Education of public in traffic laws.
- Enforcement of Regulations:
  - (a) Speed control
  - (b) Traffic control
  - (c) Training and supervision
  - (d) Observance of law and regulation
  - (e) Medical check-up
  - (f) Education of Public in Traffic Laws
- Accident Report might be useful for subsequent analys evaluation of accident cost, further legal procedures claims for compensation insurance etc.
- Accidents records are maintained with all details related to accidents.
- Collision diagram give the judgement of accident location.
- Collision diagram gives the information about other factors related to accidents like:
  - (a) Time of day
  - (b) Day of week
  - (c) Climatic conditions
  - (d) Pavement conditions
- A condition diagram is a drawing of the accident location drawn to scale.
- Scale 1 : 100 to 1 : 250 is used for drawing of condition diagram.

Practice Questions

1. Give objectives of Road accidents.
2. Give classification of Road accidents.
3. Enlist causes of accidents.
4. Give preventive measures to prevent accidents.
5. Write a note on 'Accident Report' and 'Accident Record'.
6. Draw collision diagram.
7. Give use of collision diagram.
8. Draw condition diagram.
9. Draw any four symbols used in condition diagram.
10. Give key points of condition diagram.

☆☆☆